Controversies in Sociology
edited by
Professor T. B. Bottomore and Dr M. J. Mulkay

8
Science
and the
Sociology
of
Knowledge

Controversies in Sociology

Science
and the
Sociology
of
Knowledge

MICHAEL MULKAY

Reader in Sociology, University of York

London
GEORGE ALLEN & UNWIN
Boston   Sidney

First published in 1979

GEORGE ALLEN & UNWIN LTD
40 Museum Street, London WC1A 1LU

© George Allen & Unwin (Publishers) Ltd, 1979

British Library Cataloguing in Publication Data

Mulkay, Michael Joseph
Science and the sociology of knowledge
(Controversies in sociology).
1. Science—Social aspects
I. Title   II. Series
301.24'3   Q175.5   78-40852

ISBN 0-04-301093-8
ISBN 0-04-301094-6   Pbk

Typeset in 10 on 11 point Times by Red Lion Setters, London
and printed in Great Britain
by Biddles Ltd, Guildford, Surrey

Contents

Acknowledgements

I wish to thank Tom Bottomore and John Law for reading the first draft of this book very carefully and for making a number of helpful comments and suggestions. I would also like to thank Elizabeth Chaplin for her help in preparing the final manuscript and Michael Holdsworth for giving me a special dispensation to exceed the usual length for contributions to this series.

Possibly the world of external facts is much more fertile and plastic than we have ventured to suppose; it may be that all these cosmologies and many more analyses and classifications are genuine ways of arranging what nature offers to our understanding, and that the main condition determining our selection between them is something in us rather than something in the external world.

E. A. Burtt, *The Metaphysical Foundations of Modern Science*, p. 305

If we cannot bear the paradox of accepting that genuine knowledge may be fallible, then we must ban the term altogether from productions of the human intellect.

J. R. Ravetz, *Scientific Knowledge and Its Social Problems*, p. 236

1

The Customary Sociological View of Science

The sociology of knowledge has a long history. Its origins are sometimes traced back as far as the writings of Francis Bacon, and it certainly appears as an important element in the work of the 'founding fathers' of sociology, such as Marx, Pareto and Durkheim. It continues to be a thriving area of investigation today, although its practitioners have tended to concentrate increasingly on detailed studies of specific bodies of knowledge and belief, instead of presenting the kind of general analytical formulae favoured by their forebears. Despite this long tradition, little agreement has been achieved. The field is still characterised by a great diversity of aims and interpretative schemes. This is true to such an extent that some authors are unwilling to offer any definition at all of the sociology of knowledge (Curtis and Petras, 1970, p. 7), whilst those who risk making the attempt are forced to devise very general formulations in order to cover the entire corpus of relevant literature and encompass the full range of phenomena to be studied. Thus Merton (1957, p. 456), having noted that the term 'knowledge' must be interpreted very broadly indeed in this context as covering 'virtually the entire gamut of cultural products', writes that the sociology of knowledge 'is primarily concerned with the relations between knowledge and other existential factors in the society or culture. General and even vague as this formulation of the central purpose may be, a more specific statement will not serve to include the diverse approaches which have been developed.' Within the wide range of issues covered by such a definition, one reasonably clear distinction can be made, namely, that between popular belief and commonsense or everyday knowledge, on the one hand, and systematised, specialised knowledge, on the other hand (Berger and Luckmann, 1967). In this book, I shall be concerned exclusively with the latter, that is, with the sociological analysis of specialised knowledge.

One of the central concerns of sociologists of knowledge has been to show how specialised bodies of thought and knowledge, such as aesthetic, moral and philosophical systems, religious creeds and political principles, are influenced by the social and cultural contexts in which they are produced. The guiding assumption behind this kind of analysis is clearly expressed by Mannheim. It is that the sociology of knowledge explores the 'varying ways in which objects present themselves to the subject according to differences in social settings... when and where social structures come to express themselves in the structure of assertions, and in what sense the former concretely determine the latter' (1936, p. 265). This general statement leads immediately to a number of more specific questions. We are led to ask, for instance: what kinds of social and cultural factors exercise an influence on mental productions and in what degree? What kinds of connections are there between social and cultural influences and mental productions? Which aspects of these mental productions are we trying to account for—their form, their content, their incidence, their generation or their acceptance? And most important of all, for the purposes of this chapter, which mental productions are open to this kind of sociological analysis? Are we to include all cultural products or only certain classes of them?

When we examine which areas of knowledge have actually been subject to empirical investigation, we find that scientific and mathematical thought has been almost completely ignored by sociologists. I do not mean by this that there have been no sociological studies of scientists or of the scientific community. What has been absent, until very recently, has been the empirical investigation from a sociological perspective of scientific knowledge and its social construction. In addition, although most sociologists of knowledge have discussed science in general terms, they have repeatedly rejected in principle the possibility that the form or content of scientific knowledge, as distinct from its incidence or reception, might be in some way socially contingent. Instead, they have argued strongly, albeit with occasional uncertainty, that the substance of scientific knowledge is independent of social influence and they have tried to justify this assertion on philosophical grounds. They have claimed, in short, that science is a special sociological case because it has a special epistemological status. Because this line of reasoning has been generally accepted, sociologists have left the close analysis of scientific knowledge to the philosophers of science and to the historians of ideas.

After many years in which it seemed to have been conclusively settled that scientific knowledge was exempt from sociological analysis, debate has recently begun again. This has been, in part, a consequence of radical changes in the views of science held by historians and philosophers. Throughout the 1960s, a number of historians and philosophers found themselves either bordering on or actively engaged in sociological interpretation of science as they responded to the traditional problems of their own disciplines. Gradually, these new ideas have entered sociology, helping to undermine the epistemological assumptions which had virtually required the sociology of knowledge to treat science as a special case. As the restrictions imposed by the old epistemology have become weaker, so sociologists have sought to extend and modify the work of the philosophers and historians in order to produce, for the first time, a genuine sociology of scientific knowledge. In later chapters I will describe some of the recent changes in the philosophy and historiography of science as well as certain parallel changes in sociological analysis. For the remainder of this chapter I will examine several major contributions to the sociology of knowledge and the sociology of science, in order to show that, despite occasional speculation and dissent, science has customarily been regarded as a special sociological case, and also to clarify the philosophical rationale which underlay this position.

THE CLASSIC VIEW OF SCIENCE: DURKHEIM AND MARX

All the major contributors in the nineteenth century to the incipient sociology of knowledge were doubtful about the possibility of including natural science within its scope. Let me illustrate this with respect to Emile Durkheim and Karl Marx. It is true that Durkheim attempted to provide a sociological account of the genesis of man's basic categories of thought and his forms of reasoning. He argued, for instance, that ideas of time and space, force and contradiction vary from one group to another and within the same group from one period to another. For Durkheim, the existence of such cultural variation showed that our basic categories and our rules of logic depend to some extent on factors that are historical and consequently social (1915, pp. 12-13). This appears to make an analysis of the cognitive content of science distinctly possible, because it seems that the conclusions of every particular intellectual community will be constrained at least partly by such factors as their cultural resources, the structure of their social group and their place in the wider society. But although Durkheim does not abandon this general position, he does modify some of its details so as to remove scientific knowledge from analytical consideration.

In the first place, he takes steps to avoid a completely relativist position, in which the social origin of categories and forms of reasoning could be seen to render them wholly arbitrary so far as their application to nature is concerned.

> From the fact that the ideas of time, space, class, cause or personality are constructed out of social elements, it is not necessary to conclude that they are devoid of all objective value. On the contrary, their social origin rather leads to the belief that they are not without foundation in the nature of things. (1915, p. 19)

Durkheim reaches this conclusion by postulating a unity between the physical and social worlds. Thus a group's conception of time, he suggests, will be derived from the social rhythms of its collective life. But these social rhythms will be linked to, and in a sense part of, more inclusive periodicities in the physical world. The physical and the social worlds constitute one overall realm of natural phenomena. Accordingly, in Durkheim's view, it follows that conceptions arising out of social regularities will be applicable to parallel regularities in the physical world. However, Durkheim's argument here is extremely general. It applies equally to all human groups. It establishes, at most, only that all socially derived categories will have some, unspecified, degree of 'object-ivity'. The problem remains of how to judge between divergent accounts of the physical world offered by the members of different social groups. Durkheim does not appear to recognise this difficulty explicitly. But it becomes clear in his account of social evolution that he can and does employ a more discriminating criterion of objectivity. Objectivity, he argues, becomes increasingly attainable as social evolution unfolds and as science replaces religion as the basis for human thought about the natural world.

> ...the essential ideas of scientific logic are of religious i.e. social origin... [but] science gives them a new elaboration; it purges them of all accidental elements; in a general way, it brings a spirit of criticism into all its doings, which religion ignores; it surrounds itself with pre-cautions to 'escape precipitation and bias', and to hold aside the passions, prejudices and all subjective influences... Having left religion, science tends to substitute itself for this latter in all that which concerns the cognitive and intellectual functions. (1915, p. 429)

Durkheim describes at a rather general level the social conditions which he believes are responsible for this fundamental transition in human thought. His central claim is that the growth in size of human societies and their progressive internal differentiation increasingly liber-ate intellectual activity from social constraint. Scientific thought is an outcome of this liberation and its conclusions are, therefore, compara-tively unaffected by direct social influences. Religious thought about the natural world, originating in cohesive, small-scale societies, was per-meated by categories and presuppositions derived from social life. But as societies become more complex and the form of social solidarity becomes less restricting, so it is increasingly possible for certain sectors of society to refine their conceptions and their techniques of observation in accordance with the actual realities of the natural world. The concepts and conclusions of science, he maintains, are increasingly adopted because they are true and not simply, as is primarily the case with religious beliefs, because they are collective (1915, p. 437).

A sociological analysis of science is possible, then, for Durkheim; but in a more limited form than is the case for other areas of intellectual endeavour. In principle, we can show how certain social developments have brought about the emergence of science; we can investigate whether the scientific community has certain distinctive features which make possible the institutionalisation of the scientific method and the virtual elimination of bias, prejudice and intellectual distortion; and we can observe how the minority views of the scientific specialist are received by the other sectors of highly differentiated societies. All this we can do, and perhaps more. But we cannot give a sociological account of scientific knowledge because, to the extent that it is truly scientific, it is independent of its social context. Genuine sciences, such as astronomy, physics and biology, are based on observable facts about the physical world. The conclusions of these sciences are derived from the facts, instead of being imposed upon them. Science represents phenomena not in terms of culturally contingent ideas, 'but in terms of their inherent properties' (1938, p. 35).

Durkheim's analysis of the social origins of knowledge and belief was undertaken as an explicit exercise in the sociology of knowledge. As a result, his verdict with respect to science is relatively unambiguous. In contrast, Marx's view of science as a social phenomenon emerges in a piecemeal fashion in the course of his wide-ranging examination of consciousness, ideology and modes of production. His conclusions about science are, therefore, less clear-cut and there have been somewhat different interpretations of how far he saw social factors as determining the content of science. Consequently, it will be necessary to look briefly at two different approaches, both of which claim to be forms of Marxist analysis. (There are numerous important writings on science in the Marxist tradition which I have too little space to examine here: for example, Marcuse, 1962; Habermas, 1972.) Let me begin with those relevant features of Marx's work about which there is little disagreement.

The history of mankind takes place within the natural setting provided by the objective world, a setting which is continuously transformed by human actions. By acting on the natural world man produces the means of his own existence. The repetitive relationships between people which grow out of this productive, economic activity are fundamental to all societies, constituting the major influence upon their overall structures and their mental productions. In the course of acting upon the natural world man generates knowledge about that world. This knowledge is formulated in response to the interests and economic concerns of various social groups; and is constrained by the ideological assumptions current within particular modes of production. Such knowledge is used to manipulate natural phenomena and to support, or, in certain circumstances, to transform existing social relationships.

The growth and elaboration of scientific knowledge of the natural world was greatly stimulated by the emergence of capitalist society. The economic tasks faced by the bourgeoisie in the seventeenth and eighteenth centuries drew attention to certain technological issues which stimulated an increasingly practical approach to questions of natural philosophy on the part of the intellectual representatives of this class. In due course, the new natural philosophy began to generate practically effective scientific knowledge, which was used by the capitalist class as a direct means of economic production. As scientific knowledge improved production, so more resources were made available for the support of scientific investigation. Consequently, throughout the nineteenth century and up to the present day, science has become intimately bound up with the capitalist economy and the continuous technological innovation which capitalism appears to require.

> The bourgeoisie . . . has created more massive and more colossal forces than have all preceding generations altogether. Subjection of Nature's forces to man, machinery, application of chemistry to industry and agriculture, steam navigation, electric telegraphs . . . what early century had even a presentiment that such productive forces slumbered in the lap of social labour? (Marx and Engels, 1965, p. 47)

Thus capitalism needs and promotes 'the development of the natural sciences to their highest point' (Marx, 1973, p. 409).

Initially natural science, like capitalism itself, was a liberating force, setting men free from superstition and the ideological confusions of religious thought. But in due course science necessarily became an exploitative resource for the bourgeoisie. Particularly within the realm of industrial production, science contributed significantly to the 'dehumanisation of man' (Marx, 1974, p. 97). Objective, scientific knowledge was increasingly used to create economic and administrative technologies which restricted narrowly the actions and initiatives available to their 'operatives'.

> The unity of thought and action, conception and execution, hand and mind, which capitalism threatened from its beginnings, is now attacked by a systematic dissolution employing all the resources of science and the various engineering disciplines based upon it. The subjective factor of the labor process is removed to a place among its inanimate objective factors. To the materials and instruments of production are added a 'labor force', another 'factor of production' . . . This is the ideal toward which management . . . uses and shapes every productive innovation furnished by science. (Braverman, 1974, pp. 171-2)

The central theme, then, of the Marxist analysis of science is to see the latter as a social creation and to stress that its consequences, its uses and the direction in which it develops can only be understood in relation to the wider social context. There are some similarities with Durkheim; science is viewed from an evolutionary perspective, as being set in motion

by changes in the structure of society and as undermining the effective-
ness of the religious beliefs which helped to hold together pre-capitalist
social forms. But Marx goes farther than Durkheim, although by no
means far enough, toward analysing the *production* of science in
complex, differentiated societies. This is possible for Marx because he
avoids Durkheim's reliance on an inherently ambiguous notion of simple
'correspondence' between concepts and thought, on the one hand, and
general features of social structure, on the other. Marx offers in addition
a dynamic account of social processes which can be used to describe some
of the links between science and society. In particular, he stresses that
societies are composed of relatively distinct groupings, the members of
which have opposing interests as well as an unequal capacity for
controlling the actions of others. Consequently, the direction taken by
modern science, its rapid rate of growth and the manner of its application in industry and government can be seen to have been largely
determined by the technological objectives of a particular dominant
group, namely, the bourgeoisie. The bourgeoisie has been the one group
in capitalist society able to deploy surplus economic product to generate
new scientific knowledge directly relevant to its own objectives.

But what of the form and content of scientific knowledge? In the
following passage, Marx seems to come close to arguing that the very
laws of natural science are merely a device for achieving socially
contingent objectives. Under capitalism, he claims:

> ... nature becomes purely an object for mankind, purely a matter of
> utility; *ceases to be recognised as a power for itself; and the theoretical
> discovery of autonomous laws appears merely as a ruse* so as to
> subjugate it under human needs, whether as an object of consumption
> or as a means of production. (1973, p. 410, emphasis added)

This reading of Marx has been developed most fully within the Russian
Marxist tradition. For example, in a now famous paper, Boris Hessen
(1931) tries to interpret Newton's *Principia* within a Marxist framework.
He presents evidence to show, first of all, that there was a close identity
between the central technical problems facing the entrepreneurs of the
emergent capitalist economy during Newton's period and the major
scientific problems formulated by natural philosophers at that time (see
also Merton, 1936). He also tries to show that these same technical
problems provided the focus of Newton's influential work. Consequently, Newton's work can be seen partly as an indirect response from
the intellectuals of the bourgeois class to difficulties arising in the course
of economic production. However, the *content* of Newton's *Principia*
cannot be explained quite as simply as that. Although the economic
factor is fundamental to the materialist conception of history, this does
not mean in Hessen's view that it is the sole determining influence upon

any particular set of ideas. Accordingly, he attempts to complete his
analysis of Newton's work by showing how Newton drew selectively
upon the cultural resources available to a member of his class, for
example, in the form of political, juridical, philosophical and religious
beliefs, and by showing how these ideological elements influenced and
limited Newton's thought.

Although Hessen strongly urges that science is *not* 'a passive, contem-
plative acceptance of reality, but . . . a means to effect its active recon-
struction', he manages to reconcile this view with commitment to the
Marxist-Leninist notion 'that genuine scientific knowledge of the laws of
the historical process leads with irrefutable iron necessity' to certain
political conclusions (Hessen, 1931, p. 211). Hessen's position is clearly
not without ambiguity or irony. However, I do not intend to examine
further Hessen's essay or to assess its merits and defects. It merely serves
here to illustrate that Marx can be interpreted in a strong sense, that is, as
implying that the content of established scientific knowledge should be
treated to a considerable extent as the outcome of specifiable social
processes. On the whole, however, academic sociology has not adopted
this reading of Marx. Merton, for example, having noted that Marxist
analysis allows the different spheres of mental production varying
degrees of independence from the economic base, decides that Marx and
Engels regard science as having a greater degree of independence than
any other realm of thought. The following passage is quoted as crucial
evidence for this interpretation:

> With the change of the economic foundation the entire immense super-
> structure is more or less rapidly transformed. In considering such
> transformations the distinction should always be made between the
> material transformation of the economic conditions of production
> *which can be determined with the precision of natural science*, and the
> legal, political, religious, aesthetic or philosophic—in short, ideologi-
> cal forms in which men become conscious of this conflict and fight it
> out. (Marx, 1904, p. 12, italics added by Merton)

But this passage is not viewed on its own. It is interpreted in the light of
what Merton regards as the overall trend in Marx's treatment of science.
' . . . one line of development of Marxism, from the early *German
Ideology* to the latter writings of Engels, consists in a progressive
definition (and delimitation) of the extent to which the relations of
production do in fact condition knowledge and forms of thought'
(Merton, 1973, p. 14). Merton's final reading, therefore, is that although
the focus of attention of natural science may be socially determined, this
is true neither of its conceptual apparatus nor of its substantive conclusions. Marx and Engels are seen as granting science a status quite distinct
from that of ideology.

It might be objected at this point that Merton is unlikely to be an accurate interpreter of Marx, given that his own theoretical frame of reference, that of functional analysis, is so different. Whether this is true or not, Merton's interpretation of Marx, which is also his own view, is important if only because Merton's work has exerted such a pervasive influence on the sociological analysis of science. Moreover, it is worth noting that many authors strongly committed to a Marxist analysis of science have reached similar conclusions. To illustrate this point, let me discuss briefly a recent essay by Rose and Rose. These authors certainly pay more attention than does Merton to Marx's idea that science generates its own ideology. They emphasise that 'scientism' or 'positivism' has become so dominant in present-day industrial societies that any knowledge-claim which falls outside its scope is widely regarded as necessarily vacuous. They write that:

> ... science becomes an ideology and scientists the ideologists. How does this work? As the material world controls the limits of an interpretation of the scientist in his *own* work, the answer lies, as Marx and Engels saw, *outside* the precise research area, where the scientist, freed from such constraints, talks (typically in the name of science) pure ideology. In the name of science, invoking neutrality, technique and expertise, the scientist supports the ruling strata . . . (1976, pp. 8-9)

In this passage, the authors make a distinction between the technical knowledge-claims a scientist makes within his own research network and the claims he makes in other social contexts. They suggest that the former are normally controlled by the nature of the physical world. It must be recognised, of course, that scientists will sometimes be influenced by social pressures to *propose* unjustified knowledge-claims. But, as long as there is no interference from outside with the technical criteria of adequacy applied by the research community, these socially generated claims will be judged to be inadequate by other specialists and they will be rejected. Thus within a specific research area those knowledge-claims which come to be accepted as valid can be seen to be non-ideological. They provide an accurate account of certain features of the physical world and their content is, accordingly, independent of participants' social relationships and vested interests. However, in other social contexts, the situation is quite different. Non-scientists, and indeed specialists in other areas, seldom have the technical competence to assess the adequacy of a particular scientist's claims. It follows that he will be able to use specialised knowledge so as to furnish an apparent technical rationale for policies which express his own social interests as well as the interests of other groups on whose behalf he is acting.

Claims made by scientists in the wider social context, then, will often be ideological; but this will be obscured by their technical content and by scientists' ability to invoke the 'objective facts of the natural world' as leading inevitably to certain economic, political and social conclusions. This is the kind of analysis which is generated by the Roses' Marxist approach. It is clearly different in emphasis from that of Merton and Durkheim, and directs our attention to important questions which are largely ignored by the latter. (I will return to some of these issues in the final chapter.) Yet the Roses' reading of Marx appears to coincide with that of Merton on one crucial issue. For them as for Merton, Marxist analysis recognises that scientists' knowledge-claims within their precise research areas are non-ideological.

Despite important analytical differences, therefore, the writers we have examined so far, with the possible exception of Hessen, are agreed on at least the following points: first, that science flourishes in large-scale, industrial (capitalist) societies and that within such societies, scientists create distinct communities which regulate the production of certified knowledge; secondly, that although the rate of growth, the focus of attention and the use made of scientific knowledge are in large measure socially determined, its content is independent of social influences; and thirdly, that scientific research communities are likely to have special social characteristics which reduce the impact on members' technical work of such distorting factors as bias, prejudice and irrationality, and which are, therefore, crucial in enabling scientists to generate objective knowledge. It would even be possible to bring Hessen into the fold if we were to allow a distinction to be made between the situation obtaining within capitalist society and that characteristic of socialist societies. For, as was noted above, Hessen retains the notion that ineluctable laws of nature are available within the Marxist-Leninist framework. His main thesis may, therefore, be interpreted as a claim that Newtonian science was partly pseudo-science, distorted by the social relationships of capitalism and due to be replaced by the truly scientific and determinate formulations forthcoming within socialist society (see 1931, pp. 211-12).

In the next section I shall examine the work of two more major contributors to the sociology of knowledge. This will enable me to begin to make explicit some of the philosophical assumptions which underlie the tendency to regard science as a special kind of sociological problem.

MORE RECENT VARIANTS: MANNHEIM AND STARK

Karl Mannheim is usually regarded as a central figure in the development of the sociology of knowledge (see Curtis and Petras, 1970). His work is complicated and, although his position on crucial issues undoubtedly changed as his thought evolved, he never provided a clear final statement of his framework of analysis. I will make no attempt, therefore, to give an overall account of his interpretation of the social creation of knowledge. I will instead focus narrowly on his treatment of science as a subject for sociological study.

Mannheim's sociology of knowledge includes a number of ideas taken directly from Marxism; for example, a belief in the importance of economic interests and class groupings, and in the ideological character of much social thought. In the course of his work, Mannheim tried to extend the Marxist notion of the 'existential base' to cover generations, sects and occupational groups; he also supplemented the concept of 'ideology' with an associated concept of 'utopia'; and he provided historical documentation of the connections between thought and social factors by means of several empirical studies. But the Marxist strain in his work was combined with elements taken from the German academic tradition of neo-Kantian thought (Mannheim, 1952, p. 5). One of the main ideas which he adopted from this tradition was that a radical distinction had to be made between the methods and concepts of the natural sciences, on the one hand, and those of the social sciences and historical thought, on the other. This has been widely discussed in relation to the writings of Dilthey and others (Outhwaite, 1975). I shall simply mention, therefore, a few of the points emphasised by Mannheim. In the first place, the phenomena of the material world and the relationships between them are seen as being invariant (Mannheim, 1936, p. 116). Mannheim regularly refers to the natural world, and to the concepts appropriate to its study, as being 'timeless and static'. Valid knowledge about such objective phenomena, he maintains, can be obtained only by detached, impartial observation, by reliance on sense data and by accurate measurement (Mannheim, 1952, pp. 4-16; 1936, pp. 168-9). Because the empirical relationships of the natural world are unchanging and universal, the criteria of truth by which knowledge-claims are to be judged are also permanent and uniform (1936, p. 168). It follows that natural science develops in a relatively straight line, as errors are eliminated and a growing number of truths are discerned. In short, scientific knowledge evolves through the gradual accumulation of permanently valid conclusions about a stable physical world.

Cultural products, however, cannot be investigated properly by methods of detached observation or by means of static concepts. For correct categorisation and understanding of cultural phenomena necessarily involves the interpretation of participants' meanings; and meanings cannot be simply observed like objects in the external world. Each historical period and each social group has its own distinctive values and meanings. Each analyst begins from his own culturally specific framework of meanings. Accordingly, no product of human culture can be analysed adequately from a timeless perspective. The interpretation of meanings is essentially dynamic. It must deal with the unique features of each cultural epoch and must be undertaken anew by the representatives of every succeeding historical period (Mannheim, 1952, p. 61).

Moreover, there can be no such thing as detached, uniform observation of cultural products. Their meaning must be acquired, instead, by means of involvement and sympathetic understanding (Mannheim, 1936, p. 170).

In accordance with this characterisation of the sciences and the contrasting cultural disciplines, Mannheim continually treats the advanced physical sciences as a special case from the perspective of the sociology of knowledge.

Are the existential factors in the social process merely of peripheral significance ... or do they penetrate into the 'perspective' of concrete particular assertions? ... The historical and social genesis of an idea would only be irrelevant to its ultimate validity if the temporal and social conditions of its emergence had no effect on its content and form. If this were the case, any two periods in the history of human knowledge would only be distinguished from one another by the fact that in the earlier period certain things were still unknown and certain errors still existed which, through later knowledge were completely corrected. This simple relationship ... may to a large extent be appropriate for the exact sciences ... (1936, p. 271)

In the light of several statements of this kind and in view of the importance for the whole of Mannheim's thought of the distinction between natural science and socio-historical thought, virtually all interpreters have regarded him as being entirely consistent and unambiguous in treating scientific knowledge as beyond the scope of sociological analysis (Mannheim, 1952, p. 29; Merton, 1973, p. 21; Bloor, 1976, p. 8). However, if we read Mannheim with particular care, we will observe that, on at least a few occasions, he seems to waver on this point. For example, he follows the quotation given immediately above with the following qualification: 'although indeed today the notion of the stability of the categorical structure of the exact sciences is, compared with the logic of classical physics, considerably shaken' (1936, p. 271). In this passage, Mannheim seems to be questioning his own characterisation of scientific knowledge as timeless and immutable. His uncertainty on this point becomes most noticeable when he is dealing with epistemological issues and he finally offers a resolution of the problem in epistemological terms.

The basic epistemological problem faced by Mannheim, as by Durkheim, is that of relativity. The sociology of knowledge asserts that all 'social thought', all thought outside the exact sciences, is relative to a particular social position or undertaken from a particular perspective or formulated in accordance with certain social interests. Thus in this sphere, there appear to be no generally applicable criteria for judging the validity of any specific assertion. But clearly the sociology of knowledge is itself part of this intellectual domain. It seems to follow, therefore, that there is no way of assessing the validity of its own claims—including the central claim that all social knowledge is existentially determined. Mannheim, of course, wishes to reject this conclusion and he tries to show that the assertions of the cultural sciences, although different in kind from those of the exact sciences, can still furnish true knowledge. In seeking to establish this point Mannheim is almost, but not quite, led to revise his epistemological assumptions about the natural sciences.

Mannheim does not abandon science entirely as a subject for sociological investigation and in a few pages of *Ideology and Utopia* he interprets the rise of science broadly along Marxist lines (1936, pp. 165-9). He argues that the methodology adopted by the advanced sciences was a by-product of the *Weltanschauung* of the ascendant bourgeoisie. The world view of this class, which he describes as 'democratic cosmopolitanism', denied the value of personal, qualitative 'knowledge'. Only formulations which were universally valid and necessary were allowed to stand as genuine knowledge.

> Similarly, every kind of knowledge which only certain specific historical-social groups could acquire was distrusted. Only that kind of knowledge was wanted which was free from all the influences of the subject's *Weltanschauung*. What was not noticed was that the world of the purely quantifiable and analysable was itself only discoverable on the basis of a definite *Weltanschauung*. Similarly, it was not noticed that a *Weltanschauung* is not of necessity a source of error, but often gives access to spheres of knowledge otherwise closed. (1936, p. 168)

As the bourgeoisie achieved a position of social and political preeminence, so scientific knowledge and its associated epistemology came to pervade and to dominate intellectual life. Consequently, Mannheim maintains, virtually all knowledge-claims have come to be measured against the particular epistemology derived from the dominant form of scientific knowledge.

> The particularity of the theory of knowledge holding sway today is now clearly demonstrable by the fact that the natural sciences have been selected as the ideal to which all knowledge should aspire. It is only because natural science, especially in its quantifiable phases, is largely detachable from the historical-social perspective of the investigator that the ideal of true knowledge was so construed that all attempts to attain a type of knowledge aiming at the comprehension of quality are considered as methods of inferior value. (1936, pp. 290-1)

In response to this situation, Mannheim tries to formulate an alternative epistemology which is appropriate for qualitative, existentially determined knowledge and which is in accordance with the conclusions of his own version of the sociology of knowledge.

In Mannheim's view the sociology of knowledge has shown that the 'positivist epistemology' of bourgeois society is itself partial and the product of a particular, limited *Weltanschauung*. This epistemology is, therefore, inadequate in the sense that it fails to recognise its own limitations and its own dependence on historically specific assumptions. Accordingly, its application must be confined in future to the special realm of knowledge about the physical world, to which alone it is appropriate, and it must be supplemented by a broader epistemology which recognises the partial character of all human perspectives. It has become possible for us today (Mannheim was writing in the 1930s), in a way which was impossible before the advent of the sociology of knowledge, to treat situationally detached knowledge as a marginal and special case of the situationally conditioned. From this epistemological position one assumes 'the inherently relational structure of human knowledge (just as the essentially perspectivistic nature of visually perceived objects is admitted without question)...It is not intended to assert that objects do not exist or that reliance upon observation is useless and futile but rather that the answers we get to the questions we put to the subject matter are, in certain cases, in the nature of things, possible only within the limits of the observer's perspective' (1936, p. 300). This does not mean that we are abandoning the notion of 'objectivity' or the possibility of establishing 'facts'. Rather it means that our conception of objectivity has to change.

There is no need to deny that people can often reach what they take to be 'objective' conclusions about particular phenomena, i.e. conclusions which are verifiable by the application of established procedures. Yet epistemologically these objective conclusions must be regarded as incomplete, as the product of a specific perspective and as open to revision in new social situations where other perspectives are brought to bear. In the case of different observers working within a common frame of reference, objectivity must be conceived as the application of agreed criteria of adequacy to particular knowledge-claims (1936, pp. 300-1). When participants have different perspectives, however, objectivity can only be attained in 'a more roundabout fashion'. Mannheim continues to treat objectivity as being indistinguishable from intellectual agreement, but argues that the latter will be possible only in so far as the results of each perspective are translated into the other and reconciled, usually at a more general level. This notion of divergent frames of reference becomes central to his new epistemology. (Mannheim writes of two alternative versions of his theory of knowledge, but both versions depend on this principle of resolving the differences of specific perspectives within a more comprehensive formulation.) Where some choice has to be made between perspectives, pre-eminence is given to that 'which gives evidence of the greatest comprehensiveness and the greatest fruitfulness in dealing with empirical materials' (1936, p 301). Thus the old static epistemological conception of true statements corresponding to the realities of a directly observable world has been abandoned for most realms of thought—and perhaps for thought in its entirety.

> We will have to reckon with situational determination as an inherent factor in knowledge, as well as with the theory of relationism and the theory of the changing basis of thought . . . we must reject the notion that there is a 'sphere of truth in itself' as a disruptive and unjustifiable hypothesis. *It is instructive to note that the natural sciences seem to be, in many respects, in a closely analogous situation* . . . (1936, p. 305, emphasis added)

This last sentence brings us back to the 'marginal, special case' of science. It is followed in Mannheim's text by a single page discussing certain developments in modern physics which were then quite recent. The point of the discussion is to show that the established certainties of classical physics appeared at that time to be giving way to a much more 'relativist' framework of ideas. Mannheim notes that in quantum mechanics it had come to be regarded as impossible to conceive of measurements independently of the actions and techniques involved in measuring. He points out that empirical relationships at the sub-atomic level were thought to be inherently indeterminate and that traditional notions about particles having a specific location and a definite and ascertainable trajectory of movement had now been abandoned. And, of course, he mentions Einstein's theory of relativity and the way in which it brought the position of the observer into the very equations of physics (1936, pp. 305-6).

Mannheim suggests that this trend of thought in natural science is, in its 'unformulated relationism', surprisingly similar to his own. At this point in the argument it seems possible that Mannheim will be led to reject his previous neo-Kantian portrayal of scientific knowledge as composed of universal and static truths. It seems possible that he will go on to claim that knowledge of the physical world, like that of the social, depends on the kinds of questions that we pose, on the purposes of the knowers and on their socially derived perspectives. There would be no epistemological inconsistency for Mannheim in adopting this position. For his relational epistemology provides for the 'objectivity' of socially derived knowledge-claims; and it is no less convincing to claim that active commitment to a partial perspective will reveal truths about the physical world than it is to make the same claim regarding the social world. Furthermore, this line of reasoning would have enabled Mannheim to avoid treating one major area of knowledge as a partial exception to his general epistemological principles and it would have enabled him to undertake a sociological analysis of that recent and puzzling transformation in modern physics.

Yet, despite these potential advantages, Mannheim draws back from the conclusion that scientific knowledge is in any way *socially* contingent. He does not take the decisive step of claiming that his alternative epistemology fits the physical sciences in exactly the same way as it does the historical disciplines. He is careful to state that the two realms are no more than *analogous*. Their appropriate epistemologies are parallel but distinct. The relationism of physical science is best seen as a special case of the general principle of relationism. What he seems to mean by this is that, whereas the knowledge available to observers in the socio-historical sphere is necessarily related to their social position, cultural background, group interests, and so on, the knowledge attainable by observers of the physical world is necessarily constrained only by their position in time and space. Thus Mannheim reaches a final position from which both types of knowledge are seen as inherently limited and revisable; but he maintains the distinction between the two spheres by claiming that the limitations or constraints essential to each intellectual domain are quite different in character.

To summarise, we can say that Mannheim has done three things in his discussion of epistemology. He has tried to restrict the scope of the 'positivist epistemology' to the sphere of natural science. He has tried to outline an alternative, relational epistemology for socio-historical, existentially determined thought. And he has raised the possibility that the old epistemology is not even entirely adequate for the advanced physical sciences. But he has resolved his doubts about the status of scientific knowledge without seriously challenging the orthodox epistemology and, consequently, without opening the door for a fully-fledged sociology of science. One reason why Mannheim went no further in this direction may well have been that he was entirely dependent on the work of historians of science for his own views about scientific development. He had to wait, therefore, for suitable studies in the history of ideas to emerge; and these studies did not come until long after his death. But in addition there is the fact that so much of Mannheim's thought was formulated in terms of the epistemological distinction between scientific and socio-historical knowledge. To have attempted a serious revision of the epistemology associated with natural science would have necessitated a wholesale revision of his own sociological corpus.

I have suggested, then, that some passages in Mannheim's writings contain the germ of a new philosophical view of science, one more favourable than the standard view to the possibility of a sociological approach to scientific knowledge. But these passages have generally been missed or they have been interpeted so as to make them entirely consistent with Mannheim's more frequently expressed idea that science constitutes a special epistemological, and hence sociological, case. Since

Mannheim, this latter view has become firmly entrenched in the sociology of knowledge. This is well exemplified in the writings of Werner Stark, which appeared twenty years or so after the publication of *Ideology and Utopia*. In Stark's account of the sociology of knowledge, there is no longer any questioning of the status of scientific knowledge. Mannheim's epistemological reservations have been forgotten and arguments which remained largely implicit in Mannheim are marshalled clearly and explicitly by Stark to exclude science from sociological consideration.

Stark begins by stating what he regards as a necessary assumption about the physical world, namely, that it has a fundamental permanence. There is inherent in the natural world, he suggests, a determinate state of affairs which scientists can observe and represent with increasing accuracy and completeness. Because physical scientists can achieve an abiding correspondence between an invariant nature and their own formulations, they are able to establish their distinctive degree of intellectual consensus. Scientific consensus is a product of the objectivity of scientific knowledge. This situation, however, does not apply in all so-called sciences. Like Mannheim, Stark believes that there is a fundamental difference between those concerned with the physical world and those concerned with society. 'The facts of society are made, and ever re-made, by us, whereas the facts of nature are not. They are *data* in a much more stringent meaning of the term' (Stark, 1958, p. 165).

The second major contention in Stark's analysis seems initially to be slightly at odds with the first. For, having maintained that there is a straightforward, unproblematic correspondence between true scientific knowledge and invariant natural phenomena, he suggests that another reason why knowledge of the physical world is cumulative and reliable is that such knowledge is always formulated within a perspective organised in terms of technical efficiency.

> ... whereas man has more than once shifted his vantage point for the consideration of social facts so that these facts appear to him in ever new, and often surprising, outlines, he has always kept to the same spot for surveying the facts of nature ... so that these latter facts have always offered to him the self-same surface. He has merely learned to look more closely ... Whether he likes it or not, he must, under all circumstances, pursue, among others, the economic and technological values, the value of science. (1958, p. 166)

The dependence of scientific knowledge on a particular perspective on, or attitude towards, the natural world appears at first sight to weaken the belief in the certainty and definitive character of that knowledge. Scientific knowledge seems to have become to some extent socially contingent. But Stark resolves this problem by postulating that the central value guiding man's attempts to understand the natural world has necessarily always been the same. Not only have men always sought to understand nature in order to exert control over natural processes but, in Stark's view, there is no alternative stance available to them. Thus the factual content of science is beyond the scope of sociological inquiry because it is universal. It is the product of the cumulative application of a uniform perspective to a determinate natural world.

This view of scientific knowledge has numerous further implications, some of which are illustrated by Stark. One obvious implication is that, as it is essentially the same corpus of knowledge which has been built up over time, there can be no change of meaning in the factual basis of science. True scientific knowledge can be formulated in only one way. The body of knowledge becomes increasingly comprehensive with the passage of time, but the genuine factual basis is neither revised nor reformulated. A second implication is that scientific discovery is different from that in other realms of cultural activity. In science, discovery consists not so much in creating new meanings, as it does in philosophy and the arts, as in recognising realities which, in some sense, already exist. For example: '...the astronomer merely attempts to *grasp* the data, the *pre-existent* data: he is entirely controlled by them...the scientist allows himself to be impressed by the objective truth of reality, the man of culture expresses the values in which he believes' (1958, p. 167).

Stark realises, of course, that the data of science can be conceptualised and interpreted at various levels of generality, and that certain levels and conceptions may be linked only indirectly with empirical observations. The higher levels of analysis, however, he regards as a metaphysical appendage to real scientific knowledge. The metaphysics of physics, he accepts, may well be influenced, even determined, by social factors. But the metaphysical accompaniments can be separated out from positive science. 'For science always asks: *what* is, while in the questions raised by metaphysics there always occurs the further, and disparate question of *why*?' (1958, p. 175). He gives as an example of positive, factual knowledge, the proposition that heavy bodies fall at an accelerating speed. Another example is taken from Darwinian theory, namely, the proposition that 'life is an ongoing struggle for survival in which the relatively weak are progressively eliminated and only the relatively strong survive' (1958, p. 170). This is a particularly interesting example, because the Darwinian thesis can be seen as deriving at least partly from Malthus's analysis of social dynamics. Stark argues, however, that even this is not a case where genuine scientific knowledge has been significantly moulded by social factors.

Social developments do not determine the content of scientific developments, simply because they do not determine natural facts; but they may well open the eyes of the scientists to natural facts which, though pre-existent and always there, had not been discovered before. This is what happened in the case of Darwinism ... (1958, p. 171)

Whatever its origins, the Darwinian proposition is, in Stark's view, absolutely true. It has been shown to be a summary statement of the facts of nature and is, therefore, unrevisable. The facts of nature are beyond man's control. Consequently, the content of any proposition embodying such facts cannot be determined by social factors.

Stark maintains that there is 'cause for rejoicing' in the fact that the main movements in the sociology of knowledge have agreed in all essentials with his characterisation of science (1958, p. 167), and he quotes such diverse authors as Marx, Lukács, Mannheim, Alfred Weber and Merton to illustrate this agreement. It is clear from the discussion above that there has been less certainty among sociologists of knowledge about the nature of science than Stark seems to imply. Nevertheless, I think that he is substantially correct if he is taken to mean that most writers in this field have operated within the limits of a single, standard philosophy of scientific knowledge. It is largely because sociologists of knowledge have been unable to offer a serious alternative to the standard epistemological view of science that they have been propelled into a position from which scientific knowledge and the intellectual activities of scientists have to be treated with special deference (for a discussion of 'inductivism' in the *history* of science, see Agassi, 1963). In the next section, I will outline briefly the 'standard view of science' which has been implicit in the discussion so far. In the following section I will show how it has influenced those studies which sociologists have made of the social world of science.

THE STANDARD VIEW OF SCIENTIFIC KNOWLEDGE

Most sociologists of knowledge have adopted some version or other of what Scheffler (1967) has called 'the standard view of science'. I do not mean by this that they have all endorsed every statement that I shall make below in trying to summarise the standard view. But I do mean to suggest that, although different analysts have used the standard view in different ways and with varying emphasis, sociologists' thought about science as a social phenomenon has usually been formulated within this framework of assumptions. The reader will find that all the main points contained in the following paragraphs have already been illustrated in the discussion above.

From the perspective of the standard view, the natural world is to be regarded as real and objective. Its characteristics cannot be determined by the preferences or intentions of its observers. These characteristics can, however, be more or less faithfully represented. Science is that intellectual enterprise concerned with providing an accurate account of the objects, processes and relationships occurring in the world of natural phenomena. To the extent that scientific knowledge is valid, it reveals and encapsulates in its systematic statements the true character of this world. As Galileo puts it: 'the conclusions of natural science are true and necessary, and the judgment of man has nothing to do with them' (1953, p. 63). Although the natural world is, in a certain sense, undergoing continuous change and movement, there exist underlying and unchanging uniformities. These basic empirical regularities can be expressed as universal and permanent laws of nature, which tell us what is always and everywhere the case. Unbiased, detached observation furnishes the evidence on which these laws are built. The creation of scientific knowledge 'begins with the plain and unembroidered evidence of the senses, with innocent, unprejudiced observation . . . and builds upon it a great mansion of natural law' (Medawar, 1969, p. 147). Indeed, observational laws are no more than general propositions summarising a body of reliable factual evidence. The validity of the factual foundation of scientific knowledge can be guaranteed with a high degree of confidence because science has evolved stringent criteria, for example, in connection with experimental procedures, by means of which empirical knowledge-claims are evaluated and their accurate representation of empirical phenomena is ensured. Thus accepted scientific knowledge, because it has satisfied these impersonal, technical criteria of adequacy, is independent of those subjective factors, such as personal prejudice, emotional involvement and self-interest, which might otherwise distort scientists' perception of the external world.

Although the body of scientific knowledge is basically empirical, it does contain high-level generalisations which are not observational laws and which in some cases cannot be directly derived from or tested against observations. These more abstract and more speculative propositions play an important role in scientific thought by explaining observed regularities, by co-ordinating separate observational laws into coherent intellectual frameworks and sometimes by revealing observable phenomena which were previously unknown. In certain cases, the development of new observational techniques leads to the direct confirmation of these abstract speculations and they eventually become indistinguishable from ordinary observational laws (O'Neil, 1969). But it is not necessary to conceive of theoretical laws as actually representing the realities of the natural world. It is not surprising therefore that, like statements about the 'ether', they are frequently abandoned by scientists when their usefulness has come to an end.

A fundamental distinction must be made, therefore, between observational laws and theoretical laws (Nagel, 1961, ch. 5). The latter are revisable and dispensable, but the former are not. Whilst the former deal with observable facts, the latter often deal with unobservable entities.

Nevertheless every effort is made to verify or to test theoretical laws. When a theoretical law generates inferences which are not upheld by observation, it is either revised in accordance with the new evidence or it is renounced in favour of an alternative hypothesis. In cases of uncertainty, various hypotheses will be tested until one is found which fits the full range of data. This hypothesis then becomes a candidate theoretical law. Although there is constant change and revision at the theoretical level, this is not incompatible with cumulative development at the factual level. The established facts subsumed under an abandoned theoretical law are typically passed on to its successor which, in addition, will have brought within its scope a number of newly certified facts about the natural world. 'Thus it is that science can be cumulative at the observational or experimental level, despite its lack of cumulativeness at the theoretical level. Throughout the apparent flux of changing scientific beliefs, then, there is a solid growth of knowledge which represents progress in empirical understanding' (Scheffler, 1967, p. 9).

The basic, observational laws of science are considered to be true, primary and certain, because they are built into the fabric of the natural world. Discovering a law is like discovering America, in the sense that both were already there waiting to be revealed (MacKinnon, 1972, p. 16). Once an observational law has been discovered it applies universally and it commands universal assent. There may be some slight room for cultural variation with respect to *theoretical* speculations, for their content is not wholly determined by observational data. But the greater portion of scientific knowledge, directly rooted as it is in empirical evidence, is necessarily independent of the society or the specialised group which first made it available. The social origin of scientific knowledge is almost completely irrelevant to its content, for the latter is determined by the nature of the physical world itself.

THE SOCIOLOGY OF SCIENCE

Occasionally this standard epistemology has been made fully explicit in the course of sociological investigation of science. This is done most clearly by De Gré, who states in his introduction to the sociology of science 'that a real world exists independently of our knowledge of it; that this real world is to an extent knowable through a process of approximation; and that knowledge is true to the degree to which it approximates or is isomorphic to the structure of reality' (1955, p. 37). The central implication of these assumptions, De Gré concludes, is that sociology should be concerned, not with the actual cognitive content of science, not with certified knowledge as such, but with the social conditions which make possible the attainment of objective knowledge.

Although few sociologists of science have been as explicit as De Gré about their philosophical presuppositions, they have all until recently adopted his approach to the empirical study of science. The whole tradition of sociological work on science within which De Gré writes, beginning with Merton's pioneering research in the 1930s and continuing off and on for thirty years or so, has systematically avoided examination of the substance of scientific thought (Mulkay, 1969). It has offered us 'a sociology that deals with the allegedly fixed *normative* commitments of scientists but which pays scant attention to the social significance of their patently changing *cognitive* commitments' (King, 1971, p. 15). Although this quotation conveys something important about the main tradition in the sociology of science, it would be wrong to infer from it that sociologists have concerned themselves solely with investigating the normative structure of science. Thus Merton, for instance, by far the most influential figure in this field, has studied the allocation of rewards in science as well as considering the economic, technological and military factors which facilitated the emergence and growth of modern science. Nevertheless, a consistent and continuing theme in the sociology of science has been the depiction of that 'complex of values and norms which is held to be binding on the man of science' (Merton, 1973, pp. 268-9), and which is seen as being crucially involved in the generation of certified knowledge. It is in sociologists' accounts of the supposed 'ethos of science' that epistemological presumptions become most evident and exert their clearest influence on the content of sociological analysis.

The nature of the 'scientific ethos' was first sketched out by Merton, as part of his thesis that seventeenth-century Puritanism had contributed significantly to the birth of modern science in England. Merton argued that the Puritan complex of values led to the 'largely unwitting further-ance of modern science' (1970, p. 136). Puritans emphasised cultural values such as utility, rationality, empiricism, individualism, anti-traditionalism and this-worldly asceticism. This set of inter-related values and norms was thought to parallel those characteristic of science (1970, p. 137). Consequently, the marked increase in 'scientific' activity which occurred during the seventeenth century could be seen, at least partly, as an unanticipated consequence of the increasing dominance of the Puritan movement. As a result of their religious values, those Puritans who engaged in philosophical debate tended to concern themselves with empirical issues and to address these issues in a rational, methodical and impersonal manner. (Socio-economic factors were seen as being impor-tant in focusing interest on certain kinds of empirical problems.) The fact that Puritans were over-represented amongst the adherents of the new natural philosophy and among the founding members of the Royal Society was strong evidence of there being a connection between Puritanism and the establishment of the modern scientific community.

At no point in his analysis does Merton try to establish a direct connection between Puritan values and the intellectual products of scientific endeavour. Indeed he states quite clearly in a subsequent paper that the substantive findings of science are beyond the scope of his, purely sociological, interests (1973, p. 268). His aim is, instead, the more limited one of showing that these values, in so far as they have become institutionalised in the scientific community, are essential requirements for the regular production and acceptance of properly confirmed and logically consistent statements of empirical regularities.

> The institutional goal of science is the extension of certified know-ledge. The technical methods employed toward this end provide the relevant definition of knowledge: empirically confirmed and logically consistent statements of regularities (which are, in effect, predictions). The institutional imperatives (mores) derive from the goal and the methods. The entire structure of technical and moral norms imple-ments the final objective. The technical norm of empirical evidence, adequate and reliable, is a prerequisite for sustained true prediction; the technical norm of logical consistency, a prerequisite for systematic and valid prediction. The mores of science possess a methodologic rationale but they are binding, not only because they are procedurally efficient, but because they are believed right and good. They are moral as well as technical prescriptions. Four sets of institutional imperatives —universalism, communism, disinterestedness, organised skepti-cism—are taken to comprise the ethos of modern science. (Merton, 1973, p. 270)

Scientists' belief in the goodness of these mores has its historical origins in the religious commitments of the founding members of their professional community. But these mores are also methodologically essential for the systematic creation of valid knowledge. Thus, as the scientific community has over the years severed its links with the religious sphere, scientists have ceased to justify their values in religious terms. Instead, having come to recognise the methodological import of these values, they have tended to justify them to the wider society as the cultural basis of scientific truth and as the 'pure' source of practically effective knowledge (Merton, 1970, p. xxii).

Since Merton's original formulation of the institutional imperatives of science numerous additions have been suggested: for example, norms of originality, humility, independence, emotional neutrality and impartial-ity have been proposed (Barber, 1952; Storer, 1966; Merton, 1973; Mitroff, 1974). In addition, a few critical discussions have appeared in recent years (Lemaine and Matalon, 1969; Barnes and Dolby, 1970; Weingart, 1974). But I do not intend to examine either the supplementary or the critical literature here. The point I wish to emphasise is that this portrayal of the ethos of science is powerful, and its power is demon-strated by its continued vitality over more than three decades, because it uses the standard view of science as a taken-for-granted interpretative resource. For example, if the conclusions of science *are* simply statements of observable regularities, accurate within the technical limits of the time, then it seems to follow necessarily that the particular personal or social characteristics of those proposing such statements are irrelevant to scientists' judgement of their validity. 'Objectivity precludes particularism' (Merton, 1973, p. 270). If knowledge-claims were judged on the basis of particularistic criteria, then claims would be accepted that did not correspond with the objective world. Given that valid scientific knowledge *is* objective, it follows that scientists must regularly use impersonal, universalistic criteria in the course of their professional activities. The same kind of reasoning can be applied to other elements of the scientific ethos. Organised scepticism and intellectual independence are required because scientific knowledge must not be taken on trust. All presuppositions and all knowledge-claims, including one's own, must be continually scrutinised for logical consistency and for empirical accuracy. No person's claims should be taken as valid because of his position in the scientific community. If these prescriptions were ignored, inaccurate propositions would inevitably enter the corpus of certified knowledge. Similarly the communal ownership of knowledge is required. Without free and open communication of findings it would be impossible for scientists to subject all knowledge-claims to the same critical appraisal or to apply their universalistic criteria of scientific adequacy consistently. It is not being proposed, of course, that scientists always conform in full to these principles. Deviant actions such as fraud, secrecy and intellectual prejudice do occur. But, it is argued, they occur infrequently, for otherwise natural science would not have the validity which we know it to have. To the extent that science does regularly produce valid and practically effective knowledge, it appears that these principles must have been operative.

Among the normative principles of which the 'scientific ethos' is composed, the most important is that of universalism. This principle is thought to be implemented in many different ways in science. It seems to require, for instance, that members' standing in the scientific community should be based on merit rather than any ascriptive criteria and that scientific careers should be open to all those with ability. But universalism finds its most fundamental expression in the assessment of the results of scientific research. To say that scientists judge knowledge-claims universalistically is not merely to say that scientists, like other specialists, make use of technical criteria of adequacy; for the technical criteria of different intellectual traditions or different groups may be incompatible and they may or may not serve to establish objective empirical regularities. Thus universalism means that 'truth-claims, whatever their source, are to be subjected to *pre-established impersonal criteria: consonant with observation* and with previously confirmed knowledge' (Merton, 1973,

p. 270); or, to put it another way, universalism means that the diverse strands in the development of scientific thought are guided by 'more or less common criteria and rules of evidence which transcend other differences among the contending intellectual traditions' (Merton, 1975, p. 51). On the whole, the various other elements in the normative structure are seen to contribute to the institutional goal of science by ensuring that these pre-established criteria of objectivity are rigorously applied to all knowledge-claims before the latter are accepted as certified knowledge.

This characterisation of the norms of science, intimately bound up with the standard epistemology, has strongly influenced sociologists' views of the overall structure of the scientific community as well as its relations with the wider society (Mulkay, 1977a). Let me give one example of how the standard epistemology, the established conception of scientific norms and other aspects of sociological analysis are inter-related. The assumption that scientists operate with universalistic criteria of adequacy which are sociologically unproblematic has led to a mark-edly functionalist interpretation of the system of social stratification in science (Cole and Cole, 1973). The decisive evidence on which this interpretation is based is that high rank in the research community is empirically associated with the production of high-quality results and low rank with the absence of such results; and that, as long as 'quality of results' is held constant, there seems to be no strong connection between achieved rank and other social variables. Now, if we assume that scientists judge their colleagues' work in terms of pre-established stable criteria, we are inclined to see this evidence as demonstrating that the scientific community closely resembles a meritocracy. (A lot will depend, of course, on the amount of variance in social rank which is accounted for by variations in the production of high-quality work.) It will appear that scientists undertake research and pass on their findings to their colleagues; that these findings are assessed on the basis of pre-established, impersonal criteria of adequacy and value; and that scientists are rewarded with rank in proportion to their contribution to knowledge. Thus science is one area of social life where the functional analysis of social stratification seems to work; and it is made to work because sociologists' conception of scientific knowledge allows them to assume, with no need of empirical evidence on this point, that clear, identical evaluative criteria are available to all members.

This interpretation, then, depends indirectly on the standard view of science. Once we begin to doubt whether knowledge-claims are assessed in this clear-cut fashion and to wonder whether their adequacy or quality is perhaps socially negotiable, then it becomes possible to see the system of social ranking rather differently. For example, if the criteria employed are themselves socially created and legitimated in the course of the very process of assessing knowledge-claims, the association between high rank and the production of high-quality results could have an entirely different sociological meaning. For, having acknowledged that scientists tend to establish their rank through the medium of research findings, and that there will therefore be an empirical association between quality and rank, we would still have to ask: in what ways and by what means are adequacy and value attributed to specific results? Are there perhaps systematic social differences in participants' ability to establish that their work is of high quality? Thus, although it is clear that female scientists' relatively low rank is associated with low quality of work (as recognised by other, predominantly male, scientists), we would no longer be forced to see this as the result of 'objective' differences in the findings produced by male and female researchers (Cole and Cole, 1973). It has become possible to conceive that women and the members of other social categories in science are systematically prevented from (or favoured in) establishing that their work is of high quality. By moving away from the standard view and the associated notion of 'universalism', that is, by assuming that cognitive criteria in science may be flexible and their application to particular cases problematic, it becomes possible to investigate whether the social allocation of 'quality', and thereby social rank, is affected by structural differences within the scientific community. In short, in so far as we modify the traditional epistemological belief that contributions to knowledge can be assessed objectively and unambiguously, to that extent we are able to conceive of a wider range of interpretative possibilities, *not only with respect to the social construction of scientific knowledge but also in relation to social ranking, and other social phenomena, in science*. This will become clearer in later chapters.

Clearly the major barrier preventing sociologists from exploring these possibilities is likely to be epistemological in character (Whitley, 1972). One would not hesitate to consider such possibilities with respect to 'inferior' forms of knowledge. The difficulty with respect to science is that we are departing from a well-entrenched epistemology. We are assuming that scientists' accounts of the natural world are not to be taken simply as reflections of an objective reality, nor as determined by invariant and transcendent rules of evidence. We are treating the notion of 'consonance with observation', for example, as being sociologically problematic (Collins, 1975). It is not surprising, therefore, that sociologists did not venture to pose questions of the kind suggested above until the ground had been prepared for them by a series of debates among philosophers and historians, in the course of which the customary view of science was seriously challenged.

2

Revisions of the Standard View

In this chapter I intend to describe some recent contributions to philosophy and historiography which have significant implications for the sociological analysis of science. I will make no attempt to distinguish the work of philosophers from that of historians of science: most of the thinkers whose work will be considered here bestride such disciplinary boundaries and I will refer to them simply as 'philosophers'. Nor will I try to provide a systematic exposition of each major thinker's ideas. This has been done more than adequately elsewhere (Easlea, 1973; Giddens, 1978), and I will assume that my reader is not entirely unfamiliar with the writings of such authors as Kuhn and Popper. Wherever possible, therefore, I will draw upon the work of philosophers whose writings are less well known, such as Hanson, Ravetz and Hesse; partly because they deserve to be more widely read, but also because I wish to show that the changes in the philosophy of science with which I shall be concerned are not confined to one or two well-known but unrepresentative thinkers. However, my argument in this chapter will not be organised around the views of particular writers nor in terms of the historical development of ideas about science. It will focus instead upon the central assumptions of the standard view, as they were outlined in Chapter 1. I will take four of its main contentions in turn and examine how far they need to be revised in the light of this recent body of literature.

THE UNIFORMITY OF NATURE

Both Mannheim and Stark began their attempts to establish the distinctive character of scientific knowledge with a statement of what has been called the 'principle of the uniformity of nature'. They maintained that the phenomena and relationships of the material world differ from those of the social world in being invariant and stable. In their view this is substantiated by the fact that the basic conclusions of physical science, its laws of nature, are always and everywhere the same. Although the principle of uniformity was regarded by these sociologists of knowledge as a 'necessary and crucial assumption' in any attempt to understand the nature of science, it has attracted little attention in the debates which provide the material for this chapter. Fortunately, however, the principle has been examined rather carefully by Hanson (1969). He shows that its use in the sociology of knowledge has been based on a misunderstanding and that it is a shaky foundation on which to build one's sociological endeavours.

Hanson asks the question: how do we obtain the knowledge that nature is regular, uniform and constant? How do we know that the principle of uniformity is *true*? There are only two ways in which its truth could be established, he suggests: either by formal or by empirical means. Clearly the principle of uniformity is not intended as a mere formalism. It is meant to convey something factual about the material world. Consequently, the principle of uniformity must be established empirically. But if this is so, we are faced with a vicious circularity. For if the principle is true, it must be presupposed in every empirical procedure— including that by which we hope to prove the truth of the principle itself. If the principle is not assumed, we cannot establish its validity by generalising from empirical evidence. Unless the principle is assumed to operate, there is no way in which we can infer it from particular observations.

> Hence, to gain knowledge of the truth of these principles [Hanson is discussing the principle of induction as well as the principle of uniformity] by experiment and observation is to presuppose in the search the very existence of that for which we are in search. If the principles are true, we cannot learn of it empirically, for the essence of the principles is that their truth is presupposed in every empirical enquiry. (1969, p. 408)

Hanson goes on to show with various specific examples that the principle of uniformity is vacuous in the sense that it makes no definite assertion at all. Let me give just one of his illustrations. Consider Newton's inverse-square law of gravitation and its derivative that freely falling bodies accelerate towards the Earth at 32.2 ft/sec.^{2}. Both these propositions might be called 'laws of nature'. Hanson considers whether it makes sense to ask if these laws could, under certain circumstances, be different. For example, could the law of gravitation be different on Mars, thereby overthrowing the principle of uniformity? Clearly the formula expressing the rate of free fall would be different on Mars, because this formula is derived from the inverse-square law in accordance with the special conditions associated with the Earth alone. Thus the principle of uniformity is in no way threatened by a different rate of free fall on Mars. The scope of this particular law of acceleration is necessarily restricted exclusively to the Earth. In this sense it is not a fully-fledged law of nature and our question is inappropriate. But what of the law of gravitation proper? Could we say 'The law of gravitation may take a different form on Mars' as we can say 'Freely falling bodies accelerate at a different rate on Mars'? Hanson maintains, rightly I think, that we could not. For if the 'law of gravitation' did not apply on Mars, it could no longer be regarded as a law of nature. Any *conclusive* observations, on Mars or elsewhere, which we believed to be inconsistent with the law, would not reveal an area of non-uniformity; they would show instead that our theory was wrong and our 'law' not a law at all. 'Thus, that a scientist expresses his law of gravitation in an identical form on all occasions proves nothing at all about the uniformity of nature. What it may be said to prove is that we accord the title "law of nature" to nothing which is not expressed in an identical form on all occasions' (1969, p. 353).

It seems, then, that modern philosophical analysis reveals the principle of uniformity to be no more than a rather misleading formulation of what is meant by the term 'law of nature'. The principle of uniformity is not an aspect of the natural world, but rather an aspect of scientists' methods for constructing their accounts of that world. It cannot be used, therefore, as grounds for treating the generalisations of natural science as definitive representations of a stable and uniform physical reality.

FACT AND THEORY

Belief in the inherent stability and uniformity of the physical world has often been linked to a particular view of the relationship between fact and theory in science. From this orthodox position it is assumed that certain objects and processes exist in the physical world, that certain events occur consistently and that certain stable relationships persist: these objects, processes, events and relationships constitute the facts which science has to describe accurately and explain convincingly. (For the purposes of my argument, there is no need systematically to distinguish particular facts and specific observations from general relationships between observable phenomena and the empirical generalisations which express such relationships.) These facts are seen as being theoretically neutral. They can, therefore, be expressed in a language which is independent of theory and formulated in a way which simply represents the observable realities of the physical world. Once firmly established, facts remain unaffected by interpretative advances. Indeed, as long as there have been no observational errors, they can undergo no change of content or meaning and they can be used, therefore, to discriminate objectively between theoretical alternatives.

It is accepted, of course, that successful theories usually generate new observations and new facts about the physical world. However, once such facts have been properly confirmed, they are thought to acquire an intellectual autonomy which enables them to remain unaffected even by the wholesale refutation of the analytical scheme which gave them birth.

...an experimental law, unlike a theoretical statement, invariably possesses a determinate empirical content which in principle can always be controlled by observational evidence ... even when an experimental law is explained by a given theory and is thus incorporated into the framework of the latter's ideas ... two characteristics continue to hold for the law. It retains a meaning that can be formulated independently of the theory, and it is based on observational evidence that may enable the law to survive the eventual demise of the theory. (Nagel, 1961, pp. 83-6)

This two-tier view of scientific knowledge, taken for granted by most sociologists of knowledge, has given rise to a number of intractable philosophical difficulties. In seeking to resolve these difficulties, philosophers have gradually evolved a new account of the relationship between fact and theory which has major implications for the scope of sociological analysis.

It has sometimes been maintained that when we conceive of scientific knowledge as composed of separate theoretical and factual propositions we are making use of a distinction between observable objects and unobservable or theoretical objects. Factual propositions, it is suggested, deal with the relations between observable objects, which are explained by statements involving such unobservable objects as electrons, quarks and genes. The nature of observable objects is ascertainable by direct experience and, assuming that proper experimental precautions are taken, can be established with great confidence. But the character of theoretical objects is known only indirectly and must, therefore, be treated as inherently speculative. Theoretical 'objects' are perhaps no more than convenient fictions. Electrons, for example, are to be regarded as hypothetical constructions derived from our experiences with real objects like cathode ray tubes and galvanometers. Accordingly, the notion of 'electron' can be abandoned if it proves unfruitful in the long run or incompatible with newly observed facts; and the propositions in which it features can be recast. In contrast, we can hardly deny the existence of such objects as cathode ray tubes or galvanometers; and our direct knowledge of these objects, although it can always be made more precise, is not open to doubt or repudiation to anything like the same extent.

Although this argument makes a plausible appeal to common sense, it is certainly incomplete. For instance, it deals only with *objects* and fails to cover the referents of scientific concepts concerned with relationships or processes. But, more important, there are also strong reasons for regarding the distinction between observable and unobservable objects as untenable. The central assumption is that we can distinguish unambiguously between observing an object directly and merely inferring its characteristics from its effects. But seeing an object directly is thought to involve photons moving from the object in question and impinging on the retina of the observer. As soon as we think in these terms, the notion of 'direct observation' starts to lose its clarity and, thereby, its usefulness in separating fact from theory. The vagueness of the distinction is also highlighted by considering how little is the difference between seeing directly and observing with a magnifying glass, and between the latter and using a small telescope (Smart, 1973). Yet we are required by this thesis to regard the objects revealed by the telescope as different in kind from those visible to the naked eye. Furthermore, we cannot avoid the problem simply by classifying 'seeing through a telescope' as direct observation. For we are then faced with such anomalies as the failure of those scholastics who peered through Galileo's telescope to see the objects which were so obvious to the disciple of Copernicus. We will return to such issues in the next section. For the moment we need only note that no clear distinction has been made between observable and unobservable objects; and consequently, the distinction cannot be used to strengthen the two-tier or standard view of scientific knowledge.

> ... if observability is merely a matter of degree, then there seems to be no plausible way of drawing a sharp line on this basis between objects which do and objects which do not exist. Under the influence of these considerations, most philosophers have given up the attempt to distinguish observables from unobservables on this basis and focus instead on the terminological distinction. (Grandy, 1973, p. 3)

Let us turn, therefore, to the question of whether a difference can be established between theoretical and observational *terms*.

The debate about the nature of scientific terms has been concerned particularly with the meaning of theoretical concepts. The central problem has arisen as a consequence of taking factual statements as unproblematic (except in the 'trivial' sense of depending on careful, accurate observation) and as conceptually distinct from those of theory. Thus, if statements of fact are independent of scientific theories and can be used as a neutral check upon them, and if theoretical propositions go beyond the established facts, what reference in the real world can be attributed to theoretical claims and the terms in which they are formulated? For instance, if the factual statements of geneticists about the colours of successive generations of sweet peas are quite separate from theoretical propositions about genes, about which we can obtain no direct evidence, can the latter claims be said to have scientific meaning? Furthermore, if the statements in which terms like 'gene' occur have no meaning, if they are not synthetic propositions, how can we maintain that they are true or false? And if theoretical claims are neither true nor false, they can hardly be regarded as furnishing valid knowledge.

There is some overlap between this conception of the problem and that which seeks to distinguish between kinds of scientific objects. This second formulation, however, points in a more promising direction. Whilst it appears that theoretical terms do not correspond to a special type of entity and that theoretical concepts seem to have a different kind of meaning from those used to report observations, it cannot be denied that the use of theoretical terms is peculiarly characteristic of science and that they must therefore be meaningful in *some* way.

One response to this problem is to accept that scientific theories have no direct meaning and to regard them simply as formal systems. Theoretical terms come to be seen, therefore, merely as logical devices for deriving new observational statements from established facts and as acquiring 'indirect meaning' through their linkage with factual statements (Carnap, 1939). But this answer raises further difficulties. In the first place, it is difficult to reconcile the relatively trivial role assigned to theory by this interpretation with the fact that theoretical work is regarded as of fundamental importance by practising researchers and the fact that theorists receive by far the greatest honour and respect among scientists (Hagstrom, 1965). Thus this attempt to describe the structure of scientific knowledge, although it succeeds in portraying its factual basis in a way which supports the standard view of science, does so at the cost of treating its theoretical component as largely redundant and of implying that scientists themselves are generally mistaken about the relative value of the contributions to knowledge made by theorists and observers. Clearly an account of scientific knowledge which made sense of scientists' own high regard for theory would be preferable.

A move in this direction was made with the recognition that it was in practice extremely difficult to distinguish observational terms, whose meaning was 'derived from experience', from speculative theoretical terms. For instance, there seems little point in insisting that there are two meanings of the term 'mass', one observable (as in the observed mass of a volume of gas) and one theoretical (as in the mass of its constituent molecules, which are individually unobservable yet which in sum determine the observable mass). Accordingly, it has increasingly come to be accepted that the distinction is arbitrary and as inherently ambiguous as that between observable and unobservable objects (Carnap, 1966). One major reason why it has proved so difficult to separate observational from theoretical terms is that terms seem to acquire their meaning, not as isolated units which in the case of observational terms can be referred to corresponding physical entities, but as elements within wider linguistic frameworks.

It is not correct to speak, as is often done, of 'the experiential meaning' of a term or a sentence in isolation. In the language of science, and for similar reasons even in pre-scientific discourse, a single statement usually has no experiential implications. A single sentence in a scientific theory does not, as a rule, entail any observation sentences;

consequences asserting the occurrence of certain observable phenomena can be derived from it only by conjoining it with a set of other, subsidiary, hypotheses. Of the latter, some will usually be observation sentences, others will be previously accepted theoretical statements. Thus, for example, the relativistic theory of the deflection of light rays in the gravitational field of the sun entails assertions about observable phenomena only if it is conjoined with a considerable body of astronomical and optical theory as well as a large number of specific statements about the instruments used in those observations of solar eclipses which serve to test the hypothesis in question. (Hempel, 1965, p. 112)

If this is so, if observational terms have no experiential meaning apart from their location in a broader conceptual and propositional scheme, it is simply impossible to identify a separate class of factual statements constituting the bedrock on which scientific knowledge is built.

The way in which particular observational terms acquire their meaning from a cluster of associated propositions and concepts has been made especially clear by Hesse (1974). She begins with the assumption that every physical situation is indefinitely complex. Each new situation is in detail different from every other. This leads to a loss of information in every application of observational terms, which leaves room for changes in classification to take place under certain circumstances. Many of the observational categories used in science, as well as in everyday life, are learned in concrete empirical contexts, where a direct association is established between selected aspects of the situation and a certain term. But complete fluency in the use of descriptive terms is not obtained by means of direct association alone. Learning a language also involves learning certain generalisations or 'laws' containing its terms. These laws are always linked together in symbolic networks. Knowledge of these laws and networks is required in order to identify proper occasions for the use of a physical predicate; such knowledge enables the user to apply the terms 'correctly' in situations other than those in which they were learned initially.

For example, a person may learn to observe and identify the planet Venus, partly by means of the 'law' that 'stars twinkle in the night sky but planets do not'. This law can then be used as a resource for observing other planets and may lead to the 'correct' identification of Mars, and so on. Of course, in applying this law, a newcomer may make what is regarded by a more experienced observer as a mistake. He may think that he is observing a planet, when what he is 'really' seeing is a star. If the more experienced observer wishes to correct this mistake, he may merely suggest that if his companion looks more carefully he will see that the celestial object in question is actually twinkling. But if his companion is not convinced thus easily, it may be necessary for the experienced observer to draw extensively upon his knowledge of astronomical parallax, upon theories of gravitation and optical transmission, and so on, in order to show that the object simply *cannot* be another planet; for it it were, the whole science of celestial mechanics would be overturned. The logic of this procedure is to take the current knowledge-system and its laws as given and explicitly to decide what it is we are observing in the light of the requirements of this system. Thus, the meaning of the observational term 'planet' is derived from its use within a network of related terms and propositions, and is not established simply by direct reference to a series of isolated empirical instances which can be identified independently of this cluster of interpretative resources. As Hesse herself emphasises, this account of scientific knowledge involves a far-reaching reinterpretation of the theory/observation distinction.

There are considerable implications for the standard view of science in this abandonment of the orthodox distinction between theoretical and observational terms. In the first place, if all terms obtain their meaning through their location in a framework of concepts and propositions, then it seems that no statement of fact is theoretically neutral. Scientists do not have access to independent findings against which to check theoretical alternatives. They can never step entirely outside their own analytical scheme, for to do so would deprive their concepts and their propositions of meaning. Thus all empirical statements are 'theory-laden' (Ryle, 1949). Furthermore it follows that, in so far as the analytical framework alters, so does the meaning of observation statements (whether particular findings or empirical generalisations) formulated within its frame of reference.

> . . . no feature in the total landscape of functioning of a descriptive predicate is exempt from modification under pressure from its surroundings. That any empirical law may be abandoned in the face of counter-examples is trite, but it becomes less trite when the functioning of every predicate is found to depend essentially on some laws or other and when it is also the case that any 'correct' situation of application— *even that in terms of which the term was originally introduced*—may become incorrect in order to preserve a system of laws and other applications. It is in this sense that I shall understand the 'theory dependence' or 'theory ladenness' of all descriptive predicates. (Hesse, 1974, p. 11)

We appear, then, to have reached a conclusion which rejects two basic assumptions of the standard view; that is, we have concluded that the factual claims of science are neither independent of theory nor stable in meaning. Even when the symbols on the page of a scientific textbook remain unchanged over a fairly long period, their meaning in the eyes of the research community may well be in continual flux, as the interpretative

context of research evolves. In addition, it follows that the meaning of
given factual statements will often differ for different sections of the
scientific community; for instance, for researchers as against school-
teachers and for members of separate specialties, depending on how far
these social groupings operate with divergent interpretative frameworks.
Thus not only is the factual 'basis' of science theory-dependent and
revisable in meaning, but it also appears to be socially variable.

This revised perspective on the relations between fact and theory has
several strong interpretations (e.g. Feyerabend, 1975), as well as numerous weaker versions (e.g. Scheffler, 1963). Before I discuss in more detail
the characteristics of these strong and weak versions, let me say a few
more words about the ways in which linguistic frameworks may affect
the formulation of factual propositions. There is, as I have stressed, an
influential empiricist tradition which treats facts as things or events out
there to be observed and appropriately described. Thus facts are thought
to exist, even though we may not have words with which to express them.
In order to appreciate the limitations of this view it is useful to ask, as
Hanson (1969) does: what do the facts 'out there' look (or sound or feel)
like? Could one photograph a fact? It is evident that a photograph, no
matter how clearly focused, cannot present us with facts until we begin to
select from it certain elements and to formulate them in linguistic form.
(The argument here parallels that above taken from Hesse.) The facts
represented in the photograph are those features which can be so
expressed (Strawson, 1959); and what can be so expressed depends on the
linguistic and other symbolic resources available. In other words, the
nature of our language will tend to favour certain kinds of statement,
whilst prohibiting others. (Some of the work in cognitive anthropology is
worth considering here. See Frake, 1961, and Hoijer, 1964.) This
argument is as applicable to the vocabularies of science as to any other
language system.

> ... our types of notation in physics may occasionally render us insensi-
> tive to features of the material world ... I am not saying that there *are*
> aspects of ... protozoa, and subatomic entities that elude description
> in the languages available to us. My point is only that it is not logically
> impossible that there might be. And if this point is sound we can see
> that it is not logically impossible that we *might* have come to think
> about the physical world very differently from the way we actually
> think ... Given the same physical world we might have (*logically* might
> have) come to speak of it differently ... In other words, the logical and
> grammatical traits of our several scientific languages, notations and
> symbol-clusters may affect how we see the world, or what we under-
> stand to be the facts about the world. (Hanson, 1969, pp. 182-3)

Hanson stresses that it is impossible to demonstrate this thesis other than *logically*. For it follows from the thesis itself that one cannot prove it empirically by presenting a fact which is beyond the scope of our linguistic resources. It should also be noted that no claim is being advanced that linguistic frameworks alone *determine* the content of factual statements. As Scheffler points out, scientists using an identical conceptual scheme can perfectly well formulate contradictory hypotheses. Nevertheless, it is clear that linguistic frameworks in science are usually devised in conjunction with substantive models or interpretations (Bohm, 1965); and that the latter do influence the content of factual statements in a fairly direct manner.

It follows that although . . . the same categorization allows the expression of alternative hypotheses, these hypotheses will nevertheless confer alternative meanings on the categorization in question. If I can indeed formulate the denial of my hypothesis in terms of the categorization by which it is expressed, I cannot accept such denial without altering the very meaning of this categorization, for such acceptance effects a change in my language system. (Scheffler, 1967, p. 46)

Although linguistic frameworks are necessary for the statement of facts, they give no hold on the external world until they are used to formulate some positive account of certain facets of that world. Only when this has been done can the observer attribute significance to his observations and thereby state them as meaningful propositions. This is nicely illustrated by Darwin's failure to recognise that one of his experiments had revealed what would now be seen as an important empirical regularity.

Having crossed snapdragons and produced hybrid varieties, Darwin found what he called 'prepotency'—and what Mendel called 'dominance'—in the first generation offspring. What is more he obtained both parental types in the second generation of hybrids, actually counted the number of each kind, and found 88 of the prepotent type, 37 of the other. This result is not significantly different from a Mendelian 3:1 ratio, but Darwin did not know how to attribute meaning to it. (Glass, 1953, p. 152)

Mendel himself was able to transform similar findings into a statement of empirical regularity, at least partly because he had clear expectations in mind which enabled him to regard the approximate numbers occurring in actual experiments as crude expressions of an idealised theoretical relationship (Fisher, 1936). Like Darwin, however, other well-informed and competent specialists, such as Focke, Hoffman and Nägeli, passed over Mendel's findings without seeing that they revealed 'facts of nature' not already well established. This became accepted only some forty years later, when Mendel's observations had become explicable in terms of subsequent developments in the theory of chromosomes and particulate inheritance.

So far in this section we have seen that the traditional distinctions between observable and unobservable entities, and between factual and theoretical statements are impossible to maintain without considerable qualification. The most we can do is to distinguish roughly between propositions which are closer to particular evidence and those which are used in a more general sense. But all factual claims, whether very specific statements such as Mendel's description of the results of particular crops of peas, or relatively general propositions, such as those expressing the ratios of inherited characteristics, are theory-laden. Technical terms like 'electron', 'quasar', 'gene', and the more commonsense terms and propositions which scientists regularly use (Elliot, 1974), acquire their scientific meaning from the linguistic, theoretical (and perhaps social) context in which they are embedded. The meaning of scientific observational categories and factual claims must be conceived in terms of the position they occupy in a theory. As Darwin succinctly expressed it: every fact is a fact for or against a theory (Hanson, 1969, pp. 216-17). This interpretation of the relationship between fact and theory in science, unlike the standard view, has the advantage of being consistent with the great importance that scientists themselves attach to theoretical work. It also has major implications for the sociology of knowledge, as Nagel has noted.

The import of every observation statement is therefore determined by some theory that is accepted by the investigator, so that the adequacy of a theory cannot be judged in the light of theory-neutral observation statements. Accordingly, if these claims are sound, they apparently lead to a far-reaching 'relativism of knowledge', to a scepticism concerning the possibility of achieving warranted knowledge of nature that is much more radical than the relativism associated with the views of Karl Mannheim and other sociologists of knowledge. (Nagel *et al.*, 1971, p. 18)

It is hardly surprising, therefore, that this account of scientific knowledge, certainly in its stronger forms, generates philosophical difficulties similar to those faced by Mannheim.

First, in some versions, scientific knowledge comes close to appearing empirically vacuous. The idea that factual reports are formulated in terms of their associated theoretical or metaphysical presuppositions seems to imply that each theory can only be tested in its own terms, yet that within its own terms it is immune from refutation. Any particular observation will have no meaning until it has been theoretically interpreted. However, the meaning supplied by the theory can only serve to confirm that theory. It seems only too easy to dismiss any observation which appears to depart from expectation by asserting, perhaps in a more refined form, that: 'If it departs from expectation, then it is either an incorrect observation or it is an observation of the wrong type.' Within this view of science, although an analytical framework may generate such severe internal inconsistencies that its adherents adopt some alternative scheme, it can never be unequivocally refuted by means of empirical evidence (Kuhn, 1962; Lakatos, 1970).

In addition, there is the related problem of 'incommensurability'. If the meaning of scientific terms and propositions depends on the whole belief system within which they are presented, then it is difficult to see how any two theories can be regarded as rivals or their factual claims as incompatible. Scientists working within different schemes will be investigating different worlds; and although their statements may sometimes appear superficially to be identical when taken in isolation, their meaning within their divergent frames of reference will not actually be the same. Furthermore, genuine communication will be possible only among those who share a common framework, within which given statements are understood in the same way. Kuhn (1970) has argued that these interpretative frameworks are community-based and that inability to communicate effectively is experienced only intermittently, whenever a research community's paradigm disintegrates under the pressure of internally generated inconsistencies. It is, however, just as plausible to argue that each individual scientist develops a unique perspective on his research area (Ravetz, 1971; Gilbert, 1976) and that, if the meaning of particular assertions depends on the whole framework of belief, each scientist must be 'trapped in the web of his own meanings' (Scheffler, 1967, p. 46).

This kind of difficulty closely resembles that posed by Mannheim in his attempt to devise a new epistemology for social knowledge. Like Kuhn, Mannheim assumed that agreement about facts can usually be reached fairly easily by those operating within a shared framework of meanings. But when divergent perspectives are involved between which some choice must be made, there is in Mannheim's view no alternative but to try to translate the 'opposing' perspectives to a higher level of common meaning, in such a way that the greatest fruitfulness in dealing with empirical materials is achieved. The new philosophy of science goes further than Mannheim, as Nagel notes, not only in arguing that man's view of the *physical* as well as the social world depends on shared meanings, but also in stressing the difficulty of translating from one network of meanings to another and of achieving a common understanding of what is to count as 'empirical material'.

This strong alternative to the standard view, although it has been gaining ground since the 1950s, is still a minority position among philosophers of science. Indeed, many of the latter have devoted considerable effort to amending, without entirely relinquishing, the orthodox account of fact and theory in response to the major points raised by its more radical critics. It seems to me, however, that the various moderate, compromise views which have resulted still differ sufficiently from the standard view to warrant a reappraisal of the sociology of scientific knowledge.

Certain of the ideas I have been discussing have come to be generally accepted, even by those who do not wish to abandon totally the orthodox position. In particular, it is accepted 'that the sense and use of predicates employed in the sciences, including those employed to report allegedly observed matters, is determined by the general laws and rules into which these predicates enter' (Nagel, 1971, pp. 19-20). It follows that all factual statements are corrigible in principle, that their connection with the external world is problematic and mediated by theoretical presuppositions, and that their meaning is subject to change as the analytical context itself develops. All this is accepted. What is challenged is that the meaning of factual propositions is 'determined by the *totality* of laws and rules of application belonging to the corpus of assumptions of a science at a given time' (Nagel, 1971, p. 20). Scheffler makes the same point, denying that meanings are 'so interlocked that a change in any one affects all the rest within a given language system' (1967, p. 59). In other words, it is being suggested that the cohesion, the connectedness, of scientific knowledge-systems has been exaggerated; and also that it is this misapprehension which has given rise to the philosophically unacceptable assertion of incommensurability and of empirical vacuity. Once we recognise, it is argued, that each conceptual framework does not constitute a seamless web, then we can avoid being driven to these indefensible conclusions. Let me give two brief illustrations of what philosophers like Nagel and Scheffler have in mind. I shall begin with an example taken from the field of radio astronomy (Edge and Mulkay, 1976).

Pulsars are relatively small celestial objects which emit a rapid and regular radio pulse. They were discovered quite unexpectedly in 1967-8 (Woolgar, 1976a). The research which led to their discovery began some months earlier, with the aim of identifying quasars—objects which resemble pulsars only in having very small diameters and in emitting electro-magnetic radiation at radio wavelengths. The basic idea on which this research project depended was that radio waves from very small sources fluctuated quickly and irregularly, unlike those from ordinary radio galaxies. This 'scintillation' is interpreted as a distortion of the radio waves coming from small sources as the waves pass through 'plasma clouds' surrounding the sun. It is clear that the observations of scintillating radio scources which ensued were theory-laden in the sense discussed above. Meaning was assigned to them within a complex network of assumptions; for example, assumptions about the nature of radio waves and radio receivers, about the properties of 'small' celestial objects, about the physics of the sun's atmosphere, and so on. But *pulsars* were first observed because they departed from the observational expectations generated by these assumptions. For instance, the pulses of radio emission were remarkably regular and they scintillated at night, when the effect of the sun on radio waves was minimal.

Once these unusual features had been noticed, the observers concerned explored several quite distinct interpretations of their findings— including the possibility that the pulses were artificially created by some alien intelligence. But these interpretative explorations were linked only in the most indirect manner to the original set of assumptions which had given meaning to the search for quasars. In the radio astronomers' view, the established ideas about radio waves, radio receivers, plasma clouds, and so on, were in no way threatened by this unexpected discovery. They were sufficiently indeterminate to cope with the existence of the unanticipated objects, without requiring any obvious revision. In this sense, they provided part of the background of interpretative resources within which pulsars had to be understood. However, these original assumptions were of no positive use in accounting for the unusual features of the observed radio signals. And, in fact, the observers eventually made use of a largely unnoticed paper on neutron stars which had been in the literature for some time, in order to provide a preliminary interpretation which they regarded as suitable for their findings.

This example seems to show that a single group of scientists can have access to various bodies of theoretical resources which are not closely connected; and that factual statements, although undoubtedly theory-laden, do not derive their meaning exclusively from that framework of assumptions which gave them birth. Similar conclusions are reached by Nagel in the course of an examination of some of Newton's researches into optics. Nagel notes that Newton's observational terms acquired at least part of their meaning from various laws in which they were embedded and which Newton took for granted. But, he argues, the conceptions used by Newton in interpreting his findings were in no way assumed in the design and execution of his experiments.

> [This] shows that an experiment intended to ascertain on what factors the occurrence of a certain phenomenon depends can be described so that the statement of the observations made is neutral as between alternative theories which may be proposed to explain the phenomenon, even though the descriptive statement will indeed presuppose various theories, laws, and other background information that are not in dispute in the given inquiry. (Nagel, 1971, p. 26)

Thus once we allow that the networks of assumptions and concepts used in scientific research are neither completely determinate in meaning nor wholly unified, it becomes possible to escape from the 'stultifying circularity' according to which factual statements are linked indissolubly to a particular framework of preconceptions. It does seem possible for scientists sometimes to devise independent checks for specific factual claims; although it is clear that this can only be done by treating other sets of assumptions as unproblematic, at least for the moment. Similarly, it may well be that the problem of incommensurability has been exaggerated. The examples given above, for instance, of Newtonian optics and the interpretation of pulsars, seem to show that scientists can agree about the theoretical basis for a given set of empirical results, without thereby committing themselves necessarily to identical interpretations of those results. If this is so, then at least some degree of communication is possible between participants operating with divergent analytical frameworks (Grandy, 1973). This is what Kuhn, in reflecting on his critics, has come to call 'partial communication' (Lakatos and Musgrave, 1970).

I do not intend to pursue these issues further here. I simply want to establish that, in abandoning the standard view, one does not necessarily have to adopt an extreme interpretation of scientific thought as composed of closed, self-perpetuating meaning-systems. Even the more moderate analyses of fact and theory that I have described above depart significantly from the standard view. Gone is the simple notion that science is built upon a growing corpus of neutral facts. Gone also is the idea that well-established facts are unrevisable and that, consequently, scientific knowledge accumulates in a relatively straightforward fashion. Nevertheless, although the notion of 'fact' has been made more provisional and although facts must be seen in relation to specific intellectual frameworks, it would be wrong to conceive of scientists generally as treating their observational or theoretical knowledge as merely hypothetical and in constant danger of collapse. In fact, one important factor contributing to the impressive intellectual development of modern science has been the capacity of its adherents to forget their background assumptions and to concentrate on *using* these assumptions to undertake detailed empirical exploration. Kuhn is surely correct in stressing that modern science is unusually free, compared with other areas of intellectual endeavour, from debates about fundamentals. Most scientific research is carried out in a context in which a whole series of assumptions are so firmly entrenched that their revision or refutation is virtually unthinkable. For instance, in the discovery of quasars (not pulsars) astronomers were faced, implicitly, with a choice between either recognising the existence of extraordinary objects 100 times more luminous than previously known radio galaxies, for which no remotely satisfactory theoretical account was available, or with revising views about 'redshifts' and stellar spectra which provided the basis for whole areas of astronomy and its techniques (Edge and Mulkay, 1976). It is clear that very few astronomers seriously considered the latter course, even after years of interpretative failure.

As we have seen, it is only against the background furnished by such well-entrenched assumptions that scientists are able to formulate meaningful and detailed hypotheses about the world. For each researcher and for each research community, only a limited number of issues are normally treated as empirically open. We must not, however, fall into the trap of trying to distinguish, by means of inherent characteristics, those analytical resources which are entrenched from those which are taken as contingent. To do this would be very similar to trying to distinguish observational statements unambiguously from theoretical statements. It must not be forgotten that the meaning of a scientific proposition varies with the intellectual context in which it is used. This is not merely to say that some propositions become more firmly entrenched as a research area matures, although this does tend to happen. It is to suggest, in addition, that specific scientific formulations can be employed in various different ways at the same point in time. Thus Hanson writes:

> ... law sentences are used sometimes to express contingent propositions, sometimes rules, recommendations, prescriptions, regulations, conventions, sometimes *a priori* propositions ... and sometimes formally analytic statements ... Few have appreciated the variety of uses to which law sentences can be put at any one time, indeed even in one experimental report. (1965, p. 98)

Hanson goes on to show in detail how an individual physicist can be seen to employ the second law of motion in a variety of ways as he moves from setting up an experiment, to constructing a machine, to defining an area of application, and so on. His conclusion is not surprising, for it follows directly from the idea that the meaning of a proposition depends on its connections with other formulations. It clearly allows for the possibility of considerable variation in the meaning of scientific propositions in relation to changes in *social* context, in so far as social and intellectual context vary together.

In this section I have shown that the orthodox account of the relation between fact and theory, the heart of the standard view of science, is untenable. In demonstrating this I have used implicitly an interpretation of scientific observation which is directly opposed to that of the standard view. Let me try to show clearly that a correct description of observation in science lends support to the argument above.

OBSERVATION IN SCIENCE

From within the standard view of science, observation is thought to be scientifically adequate in so far as such distorting influences as bias, intellectual prejudice and emotional involvement have been removed. Proper scientific observation occurs when the observer allows himself to be impressed by an objective reality. This view is consistent with the fact that natural philosophers, at the birth of modern science, eliminated from consideration secondary, subjective qualities, such as taste, smell and colour, and concentrated on 'objective, measurable attributes', such as motion and magnitude (Burtt, 1924). Thus observation in science has been seen as a plain recording of the unembroidered evidence of the senses and as being quite separate from the creation of meanings. Much recent work in the philosophy of science has been devoted to formulating an alternative account of scientific observation which is more consistent with the revised conception of the nature of scientific theory. In devising this account philosophers have drawn heavily on the conclusions of experimental psychologists about human perception in general.

One fundamental conclusion arising from work in psychology is that observation can never be as passive as the standard view requires. We never simply receive and register inputs from the external world. Instead we act upon that world so as to create a series of discernible but ever-changing cues about its characteristics; and in the very act of perception, the observer interprets these cues in terms of the cultural resources he brings to bear. For example, in artificial situations where we are required to observe solely by means of touch, we construct or infer the structure and composition of objects placed in our hands by performing a series of operations, such as pressing, turning and balancing. As a result of such active manipulation, we are generally able to place objects into categories and to produce statements about their overall shape and general attributes which go well beyond our actual contact with them. When asked to report what we have observed, we tend to describe the *inferred* structure in terms of the conventional categories available to us. We do not, and almost certainly could not, describe our complex series of hand movements, nor the fragmented tactile sensations produced by those movements. Our piecemeal explorations of the objects occur only on the fringes of awareness. We regard ourselves as having observed the relatively invariant structure which we have inferred out of a limited range of sensations.

The account just given of the sense of touch is perhaps not especially surprising. After all, human beings seldom observe by means of touch alone. But, more interestingly, our dominant visual sense appears to operate in a similar way. Vision also involves necessarily an active role on the part of the observer. It appears that what we see is constructed out of constantly changing sensations produced by a continuous series of movements of the eye and of the body (Bohm, 1965). For example, the eyeball vibrates in such a way that the retinal image is constantly shifted by a distance roughly equal to that between adjacent cells on the retina of the eye. Superimposed on this movement, the eye has a regular swing which is followed every so often by a sudden return approximately to the original point of focus. It appears that movement of this kind is essential to visual perception. For when experimental arrangements are devised which cancel out the effects of the eye's movement, subjects eventually become unable to see at all, even though clear images of the external world are still projected onto the retina. Thus, although we normally *see* a world filled by solid, permanent objects in three dimensions, we have no continuous optical register of these objects. Our eyes actually record an ever-changing sequence of momentary, two-dimensional and inverted impressions, out of which we construct the stable visual entities of everyday knowledge (Borger and Seaborn, 1966, p. 118).

Studies of the kind carried out by experimental psychologists seem to show that direct observation, whether scientific or otherwise, involves us in more than merely registering and reporting 'the unembroidered evidence of the senses'. The observer has no alternative but to embroider the evidence of his senses, for he receives from them no stable nor complete record of objects or processes in the physical world. Instead he receives, as a consequence of his own action upon that world, a series of continually changing cues about its characteristics. With the help of these cues, the observer is able to perform inferential work, often quite complex even in the everyday world, which enables him to decide what it is that he has observed.

> The perceived picture is therefore not just an image or reflection of our momentary sense impressions, but rather it is the outcome of an ever-changing construction . . . Such a construction functions in effect, as a kind of 'hypothesis' compatible with the observed invariant features of the person's over-all experience with the environment in question. (Bohm, 1965, p. 203)

The same conclusion is reached by Bruner in the course of his influential studies of perception/observation. He states clearly that we must no longer think of observation as providing us with a *representation* of the real world.

> What we generally mean when we speak of representation or veridicality is that perception is predictive in varying degrees. That is to say, the object that we see can also be felt and smelled and there will somehow be a match or congruity between what we see, feel and smell . . . Or, in still different terms, the categorical placement of the object leads to appropriate consequences in terms of later behaviour directed toward the perceived object: It appears as an apple, and indeed it keeps the doctor away if consumed once a day . . . The meaning of a thing, thus, is *the placement of an object in a network of hypothetical inference* concerning its other observable properties, its effects, and so on

...veridicality is not so much a matter of representation as it is a matter of what I shall call 'model building'. (1974, pp. 10-11)

In this quotation Bruner shifts (uneasily perhaps) between a behavioural and a linguistic definition of veridicality. In some contexts a behavioural definition seems reasonably appropriate; when one is considering perception by animals, for instance. In the case of animals, observation leads to behaviour without the mediation of words. (There are now, of course, a few experimental apes of whom this is no longer entirely true.) In human perception, however, the linguistic element is prominent; at least partly because human observation often leads to statements about what was observed, in accordance with which action is initiated. Scientific observation, in particular, is necessarily expressed in terms of words or equivalent symbols. (I have stressed the role of linguistic categories, but for discussion of a 'visual language' in science see Rudwick, 1976.) Scientific knowledge consists of propositions about the world, formulated in the conventional forms of a specific language system. Observation which is not contained in some kind of research report and not formulated in terms of general categories is irrelevant to the purposes of science (Ravetz, 1971). In contrast a representation pure and simple, say, in the visual form of a photograph or retinal image, contains in itself no linguistic component. (The argument here parallels that in the previous section.) A retinal image makes no assertions about the world. Yet what we see, what we observe, has a crucial bearing on our propositional knowledge. If seeing were a *purely* visual phenomenon, unaffected by the categories that the observer has at his disposal, nothing that we saw with our eyes would be relevant to what we know about the world. 'So our visual sensations must be cast in the form of language before they can even be considered in terms of what we know to be true. Until a visual sensation can be so considered, it is not observation...' (Hanson, 1974, p. 127). Even if one wishes to claim that human beings do sometimes register the external world without the mediation of linguistic categories, even if one regards some human perception as equivalent to that of other animals, it is difficult to maintain that this kind of observation is relevant to science. Scientific knowledge deals with what Ravetz has called 'intellectually constructed classes of things and events' (1971), that is, with general classifications which are defined by certain properties of their members. Particular things and events are classified (conceptualised) on the basis of how far they display the features of a particular class. And the observational reports which enter the body of scientific knowledge are couched in terms of these categories.

Scientific observation, then, is fundamentally dependent on language. For most purposes we can think of observation as the act of locating things and events by means of categories; and such categories acquire their meaning by implying that certain *statements* will be found to apply to that which has been observed. Thus, if a physicist sees an object as an X-ray tube, he thereby assumes that specific propositions will apply to this object; for instance, that under certain conditions fluorescence will appear around the anode at high voltages. Similarly, if an astronomer identifies an object as a pulsar he assumes that it will move round the heavens on sidereal time, that the pulses of electro-magnetic radiation which it emits will be regular and of short duration, and so on. It is clear that these verbal assertions are somehow implied in the act of observation, for if they were found subsequently not to apply, it would follow necessarily that the observed object was not *really* an X-ray tube or not *really* a pulsar. This is what Bruner means when he writes that observation involves placing an 'object' in a network of hypothetical inference. There is a clear correspondence here between Bruner's conclusion and the general position argued in the previous section. But the argument has now been extended explicitly to cover observation. Observation involves the application of categories to sense impressions. Categories, however, as we have seen, only have meaning within a network of related concepts and propositions. Consequently, observation consists in the interpretation of sense impressions in terms of a linguistic and theoretical framework.

This does not mean that scientists first obtain various kinds of sensations and then apply concepts and interepretations to them. Concepts and theories are always present in the very act of scientific observation (Harris, 1970). The famous passage from Duhem which follows illustrates how this is particularly clear in the advanced physical sciences, where observation is mediated through a complex vocabulary of symbolic resources. (It is appropriate to quote Duhem here, because he has greatly influenced such modern thinkers as Hanson and Hesse.)

Go into this laboratory; draw near this table crowded with so much apparatus: an electric battery, copper wire wrapped in silk, vessels filled with mercury, coils, a small iron bar carrying a mirror. An observer plunges the metallic stem of a rod, mounted with rubber, into small holes; the iron oscillates and, by means of the mirror tied to it, sends a beam of light over to a celluloid ruler, and the observer follows the movement of the light beam on it. There, no doubt, you have an experiment; by means of the vibration of this spot of light, this physicist minutely observes the oscillations of the piece of iron. Ask him now what he is doing. Is he going to answer: 'I am studying the oscillations of the piece of iron carrying this mirror?' No, he will tell you that he is measuring the electrical resistance of a coil. If you are astonished and ask him what meaning these words have, and what relation they have to the phenomena he has perceived and which you have at the same time perceived, he will reply that your question would require some very long explanations, and he will recommend that you take a course in electricity. (1962, p. 145)

It is clear that the trained observer and the untrained observer in this instance do not see the same things or the same events.

The interpretative constructions through which we observe the world generate expectations about the cues to be received in various types of physical setting. These expectations make us either more or less sensitive to different kinds of cues and they can be thought of as providing interpretative sets which enable us to translate expected cues fairly smoothly into firm observations (Bruner, 1974). Thus what we observe depends in large measure on what we 'know', and therefore expect, about the world around us. Observation is 'shot through with interpretation, expectation, and wish' (Scheffler, 1967, p. 22). The importance of this general point to scientific research has, of course, been argued emphatically and documented extensively by Kuhn (1962). Scientific observation in Kuhn's view, far from being an unselective and unstructured encounter with a series of unfamiliar flashes, sounds and bumps, is a precisely calculated creation of these as flashes, sounds and bumps of a particular kind. Science differs from commonsense knowledge, not in the elimination of preconceptions, but in the precision with which some of these preconceptions are formulated and the detail with which they are used to guide observation. 'Sometimes, as in a wave-length measurement, everything but the most esoteric detail of the result is known in advance, and the typical latitude of expectation is only somewhat greater' (Kuhn, 1962, p. 35).

A similar characterisation of this aspect of science was developed some time ago by Duhem. Duhem pointed out that ordinary testimony, based on the observational procedures of everyday life, can achieve a very high level of reliability. 'In a certain street of the city and near a certain hour I saw a white horse: that is what I affirm with certainty' (1962, p. 163). But this comparative certainty is attained only by restricting one's reports to relatively gross features of what was observed and by omitting the complex detail. In contrast, scientists try to deal in a precise manner with a complexity and minutiae of details which would defy description, if the scientist did not have at his service the clear and concise symbolic means of representation and measurement furnished by mathematical theory (Duhem, 1962, p. 164). Because scientific research is constructed and the resultant observations are expressed in terms of precise symbolic formulations, scientists are able to probe the physical world in intricate detail. Without the resources of a mathematically formulated theoretical language, an account of any routine experiment would fill a whole volume with the most confused, the most involved and the least comprehensible recital imaginable (see also Kuhn, 1963). Duhem's general conclusion will surprise many readers even today, some seventy years after it was first written.

The uninitiated believe that the result of a scientific experiment is distinguished from ordinary observation by a higher degree of certainty. They are mistaken, for the account of an experiment in physics does not have the immediate certainty, relatively easy to check, that ordinary, non-scientific testimony has. Though less certain than the latter, physical experiment is ahead of it in the number and precision of the details it causes us to know: therein lies its true and essential superiority. (1962, p. 163)

This is a far cry from the traditional view that science has evolved observational procedures which effectively eliminate uncertainty.

Scientific observations, then, are typically construed in terms of an established and complicated repertoire of interpretative formulations. In order to obtain precise observations and detailed distinctions, scientists necessarily take for granted a wide range of background assumptions. These are normally used as unproblematic resources for organising observation and for giving it scientific meaning. As Kuhn puts it, most of scientific research consists in fitting observations into the conceptual boxes provided by professional education. Yet Kuhn is surely wrong in claiming that in the great majority of specific areas of study there is only one cluster of concepts, only one paradigm or exemplar, available. As we noted in the previous section, there is a tendency in the literature to exaggerate the degree of intellectual rigidity or cohesion in science (Merton, 1975). It is, in fact, far from unusual to find that there are several candidate schemes in use in any particular area and that the research scientist's central dilemma is precisely that of choosing between these schemes (Lakatos, 1970). Nevertheless, Kuhn is clearly right in stressing that scientific discovery is often associated with the failure to match actual observations with symbolically generated expectations. When this happens, when what we expect to observe patently fails to occur, it seems likely that scientific observation comes closest to resembling the account embedded in the standard view of science. In such circumstances, to return to Duhem's example quoted above, the scientist may abandon the notion that he is measuring the electrical resistance of a coil and revert to describing in gross terms the oscillations of a piece of iron carrying a mirror.

> ... these observational situations have a point to them just because they contrast with our more usual cases of seeing. The language of shapes, color patches, oscillations, and pointer readings is the language appropriate to the unsettled experimental situation, where confusion and perhaps even conceptual muddlement dominate. And the *seeing* that figures in such situations is the sort where the observer *does not know what he is seeing*. He will not be satisfied until he does know, until his observations cohere and are intelligible as against the general background of his already accepted and established knowledge. And it

is this latter kind of seeing that is the goal of observation. For it is largely in terms of it, and seldom in terms of merely phenomenal seeing, that new inquiry will proceed. (Hanson, 1969, pp. 108-9)

It is important to stress, once more, that what Hanson here calls 'phenomenal seeing' is not a mere registering of the external world. Even phenomenal observation is an interpretative act, but one in which generally available cultural resources are utilised in place of specialised technical vocabulary. Thus another way of saying that the scientist *does not know what he is seeing* is to say that the formulation of his observations in the crude terminology of everyday life is normally deemed to be inadequate in the social context of physical research. However, such a formulation may well be entirely appropriate in the optician's testing room, in the psychologist's laboratory, or when theoretically generated expectations no longer appear to hold (see also Chapter 4 below). It seems, therefore, that what is to count as an accurate observation of physical phenomena will vary from one social setting to another (Lewis, 1956, p. 52).

It appears, then, that observation of the physical world has at least the following features, none of which is consistent with the standard view of science. It is an active process, in the sense that the observer creates and responds to a dynamic sequence of cues. It involves categorisation, in that cues are used to place hypothesised objects and events in terms of pre-established sets of concepts. It is inferential, that is, the observer necessarily generalises from a range of cues which is always 'incomplete' in order to establish what it is he perceives. Observation is not separate from interpretation; rather these are two facets of a single process. In most scientific research, observation is intimately guided by and expressed in terms of a complex repertoire of symbolic formulations. These give rise to comparatively precise expectations, in relation to which observations can be assigned their scientific significance. The quantified language of science as well as its controlled experimental procedures produce exceptionally detailed and diverse empirical evidence. But neither the certainty of these observations nor the correctness of their formulation is in any way guaranteed. Nor is there any 'one right way' of reporting the results of a given observation. Judgements of observational adequacy seem to vary, like the meaning of propositions, according to the interpretative and social context. This theme will be taken up again in the next section.

THE ASSESSMENT OF KNOWLEDGE-CLAIMS

The discussion so far of the principle of uniformity, the nature of observation and the interdependence of fact and theory has major implications for every other component of the standard view of science

(Easlea, 1973). Once we have departed from the traditional interpretation of fact, theory and observation, we must necessarily move towards a new account of the processes of scientific discovery, the nature of scientific consensus, the character of scientific progress, and so on. I shall not attempt to cover all of these ramifications in the rest of this chapter. Instead, I shall concentrate on examining how far we can continue to maintain that there exist common criteria and rules of evidence for assessing scientific knowledge-claims, which are applicable irrespective of differences in substantive concern or analytical approach. This issue is worth exploring, not only because the assumption that there are such criteria is an important part of the standard view, but also because it will help me to indicate some of the sociological implications of the newer philosophy of science.

In the philosophical debate about the criteria used to certify scientific knowledge, most attention has been given to the principles involved in validating *theoretical* claims. Popper is not unrepresentative in suggesting that it is scientists' way of choosing between theories which make science rational (1963, p. 215). Despite philosophers' emphasis on theory, it is clear that most practising scientists are also required, and probably more frequently, to make judgements about the accuracy or reliability of particular observational reports. It may initially seem surprising, therefore, that there has been so little concern with the criteria by which specific data are judged, especially in view of the fact that 'agreement with the data' has been so frequently proposed as a fundamental principle for the assessment of theoretical claims (Frank, 1961). However, one reason for concentrating rather more on theoretical than on observational criteria is that observation is itself theory-laden. Consequently, when general criteria are used to judge the accuracy of particular observational reports, they are interpreted in terms of specific theoretical assumptions and a specific analytical context.

Let us take, for instance, the criterion of 'replicability'. There can be no doubt that scientists often use a notion of this kind in deciding whether or not particular experimental claims should be accepted (Ravetz, 1971). Empirical results which cannot be reproduced under specified conditions are usually regarded as untrustworthy. However, the application of such a rule to specific instances is far from unproblematic; for what counts as following a rule cannot be ascertained from inspection of the rule itself (Wittgenstein, 1953). What is to count as a 'replication' depends on scientists' theories about the phenomena under study and on their view of the factors which may influence the observational situation. Consequently, as theoretical frameworks evolve and experimental techniques develop, so the way in which the general criterion of 'replicability' is applied in any given area necessarily alters. This can be illustrated by referring back to the earlier discussion of nineteenth-century work on inheritance. Before the emergence of Mendelian genetics, Darwin's experimental results on snapdragons were hardly better than random numbers; and although there was no reason to doubt the reported observations of such a renowned naturalist (see below, however, *vis-à-vis* Mendel), there was no independent way of checking their accuracy. To have repeated Darwin's experiment and obtained different numbers would have had no clear implications with respect to the latter's findings. For numbers have to be interpreted before one can judge what is to count as an equivalent result. Thus, it was not until after the theoretical developments at the end of the nineteenth century, and the accompanying advances in statistical techniques of inference, that Darwin's findings could be seen to be 'not significantly different from a Mendelian 3:1 ratio'. The acceptance of Mendel's work by geneticists, and its theoretical interpretation, furnished them with criteria in terms of which Darwin's findings could be seen to be accurate and, indeed, to provide further confirmation (or replication) of Mendel's own results. Darwin's observational accuracy (within acceptable limits) was guaranteed by its conformity with theoretical expectation.

The dependence of 'replicability' on theoretical context is revealed even more clearly in the attempt by Fisher during the 1930s to 'reconstruct' Mendel's original experiments as exactly as possible. Although he was able to confirm many of the latter's results, Fisher had some major reservations.

A serious and almost inexplicable discrepancy has, however, appeared, in that in one series of results the numbers observed agree excellently with the two to one ratio [this experiment required a 2:1 and not a 3:1 ratio], which Mendel himself expected, but differ significantly from what should have been expected had his theory been corrected to allow for the small size of his test progenies. To suppose that Mendel recognized this theoretical complication, and adjusted the frequencies supposedly observed to allow for it, would be to contravene the weight of the evidence supplied in detail by his paper as a whole. Although no explanation can be expected to be satisfactory, it remains a possibility among others that Mendel was deceived by some assistant who knew too well what was expected. This possibility is supported by independent evidence that the data of most, if not all, of the experiments have been falsified so as to agree closely with Mendel's expectations. (1936, p. 132)

Fisher's remarks show clearly how the criteria for identifying accurate observations had changed as the corpus of genetic knowledge had grown. Although geneticists were still working within a Mendelian framework, they had developed criteria for assessing experimental results on small populations (based on statistical analysis of variation in small populations) which in retrospect made some of Mendel's observational claims appear extremely unlikely. Accordingly Fisher suggests that someone had tampered with the results so as to ensure that they satisfied the criteria of accuracy which would have been assumed to apply some seventy years before.

This example also shows how an apparent failure to replicate in accordance with current observational criteria can be easily reinterpreted within a given theoretical context in such a way that established views remain secure. By the time that Fisher's work was carried out, the conclusions drawn from Mendel's experiments were too well entrenched to be easily challenged—even by the demonstration that some of the original results were suspect. What Fisher does, therefore, is to take the Mendelian framework and its subsequent elaborations for granted and to use these interpetative resources in reconstructing Mendel's research programme. He is disturbed to find that some of Mendel's supposed findings appear to have been spurious. But Fisher is forced to accept this conclusion because, in view of the current framework of analysis, *he knows what Mendel's actual results must have been like*. Consequently, he is obliged to 'explain away' apparent observations which do not conform to expectation; and he does this simply by asserting that the data must have been wrongly recorded. Fisher's tentative explanation of Mendel's suspiciously precise results in terms of a hypothetical assistant, who knew too well what was expected, illustrates with what flexibility scientists can reinterpret observational material in order to avoid inconsistency with 'established knowledge'.

When we look at all closely at the observational criterion of 'replicability' we find that it is almost a mere formality. It has no content until it is put into practice in a particular scientific context. It can perhaps best be thought of as a recommendation that scientists ought to try to produce numerous 'equivalent' observations in connection with any given problem. But replication introduces no new criteria by which experimental results can be judged. Nothing more is involved in replication than the normal procedures of interpreting findings in accordance with other evidence and in the light of analytical conceptions. If this illustration is at all representative, it seems that the kind of observational criteria discussed in the philosophical literature obtain their meaning from the theoretical or interpretative context in which they are applied. Let us turn, therefore, to examine the principles by which theories are thought to be validated.

Numerous criteria appear to be cited by scientists in assessing relatively generalised or theoretical knowledge claims; criteria such as agreement with the evidence, simplicity, accuracy, scope, fruitfulness and elegance. One of the main difficulties with these principles is that they deal with quite different dimensions. How, then, is it possible for scientists to combine them in a manner which is not arbitrary and which provides the common criteria required by the standard view? For instance, how is the requisite of accuracy to be reconciled with that of simplicity? How much inaccuracy will we allow in order to achieve a given level of simplicity, and why? Such questions are undoubtedly resolved in practice. Thus 'physics is filled with laws that express proportionality, such as Hooke's law in elasticity or Ohm's law in electrodynamics. In all these cases, there is no doubt that a non-linear relationship would describe the facts in a more accurate way, but one tries to get along with a linear law as much as possible' (Frank, 1961, p. 14). In these instances, generalisations are stated in such a way that accuracy and agreement with the evidence are to some extent sacrificed in order to achieve a convenient level of simplicity and to facilitate certain kinds of mathematical computation. But the balance between these different dimensions which is achieved in any particular set of formulations is clearly not given by the physical world. It is rather a conventional arrangement which differs over time and from one group of scientists to another (Ravetz, 1971; Lakatos, 1976). This does not mean that the ways in which criteria are combined are entirely arbitrary. But it does mean that different procedures, and therefore different knowledge-claims, would be acceptable in different interpretative contexts.

There was a time when, in physics, laws that could be expressed without using differential calculus were preferred, and in the long struggle between the corpuscular and the wave theories of light, the argument was rife that the corpuscular theory was mathematically simpler, while the wave theory required the solution of boundary problems of partial differential equations, [deemed to be] a highly complex matter. We note that even a purely mathematical estimation of simplicity depends upon the state of culture of a certain period. (Frank, 1961, p. 14)

In short, the criteria used in evaluating theoretical claims, like those applied to particular observational reports, seem to vary in meaning in accordance with the context in which they are used. They cannot be regarded, therefore, as providing a means of assessing knowledge-claims which is independent of specific analytical commitments.

This conclusion applies just as much to the criterion of 'consonance with observation' as to the other more obviously conventional criteria. There was a time when it was thought that scientific theories or generalisations were accepted when they could be proved to be true, that is, in agreement with the observable facts (Frank, 1961). Thus scientific knowledge-claims were seen as claims to have established the truth (Shepherd and Johnston, 1976). But this view has been shown to be inadequate in the course of the philosophical debate about induction. It is inadequate because, logically, it is impossible to prove any general, law-like proposition by means of evidence about particular instances. General propositions cannot be induced from a particular set of observations with certainty, because each *general* proposition necessarily goes beyond the cases already observed, to include others about which there is at present no evidence. In other words, the universal generalisations of science must be seen as being imaginatively hypothesised, or extrapolated, from essentially incomplete sets of observations. Moreover, such universal propositions can never be subsequently proved, even by means of the most extensive and detailed series of successful predictions. For there can be no complete guarantee that the *next* test will not produce such disagreement between prediction and observational results that the universality of the proposition will have to be abandoned. (This argument presupposes that observations are not completely determined by the theory they are intended to test.)

It has therefore been suggested, particularly by Popper (1959, 1963), that the major criterion of theoretical adequacy in science is and should be the ability of a claim to withstand attempts at falsification. This argument seems initially plausible because, in principle, although an immense number of positive observations is insufficient to prove a generalisation, one negative observation seems enough to disprove it. However, this thesis would only hold in relation to isolated theoretical propositions which could be compared with absolutely unproblematic observations. As soon as we take into account that any theoretical proposition is linked to and depends upon others, that observation is itself an interpretative act and that some theoretical assumptions must be made in order to establish the meaning of an observation, the appealing simplicity of the criterion of 'resistance to falsification' is lost. We are never in a position where we can measure an isolated and simple theoretical statement against an unmediated natural world. Instead we formulate and compare complex networks of theoretical and observational statements, and we seek to establish as consistent an account overall as possible of the phenomena with which we are concerned. We never, therefore, propose a theory in such a way 'that Nature may shout NO. Rather, we propose a maze of theories, and Nature may shout INCONSISTENT' (Lakatos, 1968, p. 162). The most that scientists can achieve, then, is a high level of consistency among the various components of their analytical framework.

One clear implication of this philosophical debate is that scientific knowledge is inherently inconclusive. For it is impossible to move conclusively from reports about particular samples of things and events to statements about general classes of things and events (Ravetz, 1971). It seems to follow that, because formally valid demonstrations yielding certainty are unattainable in science, scientific knowledge-claims are assessed not for their truth but for their capacity to meet the requirements of a particular interpretative context. Such contexts in science contain at least the two following general requirements: first, consistency with other knowledge-claims, which may vary in their level of generality and in the degree to which they are firmly established; secondly, conformity to conventional standards of adequacy, for example, in connection with quantitative precision, rigour of argument, range of evidence, etc., which are regarded by the members of a research 'community' to be appropriate to a given class of problems. The relevance of prior knowledge-claims to the assessment of subsequent claims has already been demonstrated sufficiently. We have also noted examples of variation in the conventional criteria of mathematical rigour and experimental accuracy. It is worth emphasising the limitations of these two elements as bases for assessing new knowledge-claims. Prior claims can give us no certainty when we use them to screen subsequent claims, because all scientific claims are inherently inconclusive. Criteria of adequacy are no more certain, for they cannot even be established by normal scientific procedures, that is, by means of argument based on controlled observation. This is partly because they are often difficult to make explicit, being akin to the tacit knowledge which is passed on by craftsmen through personal contact; a characteristic which helps to explain why philosophers have not succeeded in studying them in any detail. It also means that they are not easily subject to critical public evaluation via the journals. Thus, whilst criteria of scientific adequacy operate as resources for judging knowledge-claims, their own 'adequacy' can only be established in a most indirect and uncertain manner (Ravetz, 1971).

It is important to note that an item can appear to satisfy current criteria of adequacy as an observational or theoretical claim and yet be rejected on grounds of inconsistency with established knowledge. Consistency of this latter kind is crucial in determining whether or not a particular claim is rejected. Nevertheless, at least in mature fields, there is always a close connection between the body of established knowledge and the criteria of adequacy used by its practitioners. The range and complexity of such criteria operative in any field reflect the structure of the interpretative scheme in use. Consequently, although it is possible to distinguish certain general classes of criteria, the members of any one class are realised in different ways in each area of research. Ravetz identifies two broad types of criteria of adequacy: those relating to processes of inference and those relating to evidence. The latter are more varied than the former:

> . . . for they control not only the conditions of the production of data and information but also the strength and fit of the evidence in its particular context. It will frequently be necessary for some of the evidence to be explained and defended explicitly; and these subsidiary arguments must also meet criteria of adequacy appropriate to their function. Thus the complexity of a solved problem is matched by that of the set of relevant criteria of adequacy; and that set will depend closely on the field of inquiry. Hence *it is impossible to produce an*

explicit list of criteria of adequacy applying to a wide class of problems. (Ravetz, 1971, pp. 154-5)

It follows from this analysis that criteria of adequacy will change and evolve as the technical culture alters in which they are embedded. 'A scientific problem, unlike a textbook exercise, carries with it no guarantee that there exists a "correct" solution against which those actually achieved can be tested' (Ravetz, 1971, p. 149). (Although in a situation resembling Kuhnian normal science, the existence of such solutions may be *assumed* by participants.) Thus, any significant knowledge-claim is likely to entail some revision of criteria of adequacy, as well as involving alteration to the current theoretical framework. Let me illustrate this point with another example taken from radio astronomy (Edge and Mulkay, 1976). When the first reports of celestial radio emission were published in the 1930s, they were virtually ignored by physicists and by astronomers. There were sound reasons for this. In the first place there was Planck's Law of black-body radiation, from which it could be inferred that radio emission from celestial objects would not be detectable. Secondly, there was the apparent anomaly that the sun, despite being a dominant source of electro-magnetic radiation at visual wavelengths, was reported to be a relatively weak source of such radiation at radio wavelengths. Thirdly, the early findings in radio astronomy seemed to show that there were effectively two separate universes: the traditional one made up of optically observable objects and a previously unsuspected universe composed of an almost completely distinct collection of objects detectable only by radio techniques. As a result of these and other apparent anomalies, optical astronomers on the whole were unwilling to pay much attention to the early knowledge-claims of radio astronomers. These claims were too difficult to reconcile with well-established optical knowledge. And in addition, they did not satisfy the standards of precision or rigour required by the community of optical astronomers. Consequently it was possible for astronomers to dismiss these startling assertions until the evidence on which they were based was 'adequate', that is, comparable to the results of optical work.

In the long run, of course, optical astronomers did change their interpretation of many celestial phenomena in response to the findings of radio astronomers. Theorists were the first to respond favourably; partly because, being less involved in observation, they were less committed to the criteria of adequacy governing optical observation. They were able to recognise that this new kind of knowledge-claim had to be judged initially by different, and less demanding, standards. Close collaboration in observational work, however, occurred very seldom until radio astronomers had achieved standards which approximated to the established criteria of optical adequacy. These latter criteria had been evolved within the long tradition of optical research and were appropriate for current visual techniques. But they were quite inappropriate when applied to the crude initial observations by radio methods, at least in the view of radio astronomers. It is clear that the first knowledge-claims to be proposed by the latter were inconsistent with both the established astronomical framework of interpretation and the associated criteria of observational adequacy. For their claims to be considered, not only was it necessary to accept that the body of optical knowledge might need extensive revision, but also that the very criteria by which an astronomical knowledge-claim could be deemed worthy of investigation might have to be set aside. In cases of major claims like this, it is quite misleading to think in terms of assessment by the application of clear and predetermined criteria. Rather the knowledge-claim, if it signifies at all, implies that the cultural resources available for purposes of assessment are themselves in need of amendment.

Because the evaluation of scientific knowledge claims is highly complex, involving a subtle balance of tacit criteria of adequacy with estimates of consistency, significance and analytical import, it tends to be a somewhat prolonged process—except in cases where claims are universally ignored as trivial or unanimously rejected as inconceivable. In the course of this process, knowledge-claims move through a sequence of phases, within which their content and their meaning are continually reinterpreted in accordance with the demands of different interpretative and social contexts.

One social context through which almost all significant scientific knowledge-claims pass is that of evaluation on behalf of a professional journal. But publication in a journal in no way establishes that a claim has been accepted by the scientific community. The assessment carried out at this juncture is relatively routine. Seldom is any attempt made to reproduce experimental data; and complex arguments are not usually examined in great detail. In short, the knowledge-claim is subject at this stage to a fairly superficial and preliminary appraisal of adequacy, consistency and significance (Mulkay and Williams, 1971; Ravetz, 1971). However, a much more stringent selection procedure comes into operation after the claim has become publicly available. For it is now treated by other researchers as a symbolic resource. At this point, each interested researcher asks, in effect: 'How far does this contribution help me to produce "better" solutions to my own problem?' It seems that the answer to this question is usually negative, for the great majority of published reports receive little attention in the subsequent literature. Thus claims which are found wanting are seldom publicly refuted. They are usually ignored instead. A relatively small proportion of knowledge-claims, however, are noticed and used widely by other researchers. But the elements which are regarded by others as the most significant parts of a claim are not necessarily the same as those intended by its author; nor is the claim necessarily interpreted in the same way. As a knowledge-claim becomes divorced from its original context, it is subject to varying reinterpretations even whilst remaining within one area of investigation.

> When a solved problem has been presented to the community, and new work is done on its basis, then the objects of investigation will necessarily change, sometimes only slightly, but sometimes drastically. In a retrospect on the original problem, even after a brief period of development, its argument will be seen as concerning objects which no longer exist. There is then the question of whether it can be translated or recast so as to relate to the newer objects descended from the original ones, and still be an adequate foundation for a conclusion. If not, then the original conclusion is rejected as dealing with non-objects, or as ascribing false properties to real objects. But if such a translation or recasting is possible, then the original solved problem is seen to have contained some element which is invariant with respect to the changes in the objects of investigation. (Ravetz, 1971, p. 189)

It is this informal process of selection that directs the course of scientific development and that transforms those few formulations which withstand reinterpretation and which continue to generate 'successful' conclusions into the major intellectual achievements of science (Toulmin, 1961).

Those select few assertions which are found to be capable of wide-ranging and fruitful elaboration are passed from one research area to another and eventually reach the scientific textbook and the historical record. These influential formulations are undoubtedly part of scientific knowledge. There is, however, no clear distinction to be made between those claims which contribute to knowledge and those which do not, apart from their interpretation and use in a particular context; for a claim unambiguously rejected by one research community can be treated by another as a valid and fundamental assertion. Mendel's work once more stands as a well-known example. Furthermore, the meaning and content of these formulations does not remain stable. They undergo what Ravetz has called a 'process of standardization' (1971, pp. 199-208). Ravetz argues that a claim *must* be standardised if it is to become a fairly general component of scientific knowledge. If an assertion remains attached to its original problem, it will not outlive the solution of that problem. But in so far as it is divorced from that problem, it must be stated in some standard form which can be employed for various new purposes and in diverse scientific contexts. In this process, however, some of the original content is lost. Meaning is lost by translation in science as well as in literature. Points of obscurity and conceptual difficulties are overlooked. The limitations of underlying assumptions are forgotten. And the balance and emphasis of the original formulation are altered to meet the needs of new areas of application. In addition, because the knowledge, technical skills and standards of adequacy of the various audiences involved are likely to be quite diverse, the standardised version must be considerably simplified. This process reaches its climax in the passage of certified knowledge from the research community to the school curriculum and the popular text; but these are merely the most obvious parts of a continuous sequence of reinterpretation and change of meaning (for an empirical illustration, see Gilbert, 1976a).

> ... the content of a standardized fact may decay, almost without limit; the degree of sophistication and of faithfulness to its original which is necessary for its adequate performance of its function will depend very strongly on its use. Also, it can be seen that a version of a standardized fact which is good enough for one function can be quite inadequate for another; and since any standardized fact performs a variety of functions, it will naturally appear in a variety of versions. (Ravetz, 1971, p. 202)

The conclusions of this section can be briefly summarised as follows. Scientific knowledge-claims are not assessed by means of invariant, universal criteria. Although certain broad conceptions have been identified in the philosophical literature as common bases for accepting or rejecting claims, these conceptions are necessarily interpreted by scientists in terms of particular theoretical ideas and specific analytical repertoires. The rules of evidence, criteria of consistency, and so on, in science are not rigid. They are certainly flexible enough to allow scientists considerable leeway in interpreting evidence so as to support well-entrenched assumptions. Moreover, the standards applied in the selection of knowledge-claims are not inherent in the phenomena of the physical world. All claims are judged partly according to conventional criteria of adequacy, which vary over time and from one group or social context to another, and partly in terms of their consistency with an ever-changing interpretative framework. Any significant claim is likely to entail some revision of current criteria of adequacy, as well as implying that the established corpus of knowledge is inadequate in some way. Consequently, the assessment of such claims within a research specialty tends to occur relatively slowly, and is often characterised by marked opposition, as members explore the implications of the claim. The process of assessment is, therefore, also a process of reinterpretation. Similarly, reinterpretation or meaning-change occurs, in accordance with different criteria of adequacy and different analytical purposes, whenever assertions enter another research area or move altogether outside the research community. In short, contrary to the standard view, it seems that scientific knowledge is not stable in meaning, not independent of social context and not certified by the application of generally agreed procedures of verification.

SOCIOLOGICAL IMPLICATIONS

In so far as the arguments presented in this chapter, and in the literature on which it draws, are deemed to be valid, the epistemological barrier to a sociological analysis of scientific knowledge has been removed; although many practical difficulties still remain. Sociologists of knowledge treated science as beyond the scope of their analysis because it was thought to be a special kind of knowledge. This it clearly is in some respects, for instance, in the extent to which it has achieved intellectual supremacy in the modern world. Thus when they stressed that science and the scientific community were special *social* phenomena, thinkers like Durkheim and Mannheim were undoubtedly correct. But their view of scientific knowledge required them to place a strict limit on their conception of the social character of science. In the first place, they saw that science played an important part in modern society because the provision of objective knowledge had many social side-effects, not the least of which was the displacement of religion noted by Durkheim. Secondly, science was sociologically significant because scientists seemed not only to have devised a research method which progressively revealed the realities of the external world, but also to have evolved an appropriate form of social organisation which kept their knowledge exceptionally free from distortion by social or personal influences. It was this social aspect of science which particularly interested Merton. But sociological analysis had to stop there. Thus we find Mannheim eagerly accepting that the findings of physics are relative to the observer's position in time and space, but drawing back from the possibility that they might in some sense be *socially* relative. Sociological analysis could not go this far, for science was not simply one social construction among others; it was 'real, certain, indubitable and demonstrable knowledge' of an objective world (Popper, 1963, p. 93). Sociology could deal either with the social conditions which helped to reveal (or to hide) the objective world or with the social consequences of objective knowledge. Sociology could say nothing about the form or content of scientific knowledge itself, *because the conclusions of science were thought to be determined by the physical and not the social world.*

I have tried to show in this chapter that there are good grounds for rejecting this portrayal of science. In particular, the central assumption that scientific knowledge is based on a direct representation of the physical world has been criticised from several directions. For instance, factual statements have been shown to depend on speculative assumptions. Observation has been shown to be guided by linguistic categories. And the acceptance of knowledge-claims has been shown to involve indeterminate and variable criteria. Scientific knowledge, then, necessarily offers an account of the physical world which is mediated through available cultural resources; and these resources are in no way definitive. The indeterminacy of scientific criteria, the inconclusive character of the general knowledge-claims of science, the dependence of such claims on the available symbolic resources all indicate that the physical world could be analysed perfectly adequately by means of language and presuppositions quite different from those employed in the modern scientific community. *There is, therefore, nothing in the physical world which uniquely determines the conclusions of that community.* It is, of course, self-evident that the external world exerts constraint on the conclusions of science. But this constraint operates through the meanings created by scientists in their attempts to interpret that world. These meanings, as we have seen, are inherently inconclusive, continually revised and partly dependent on the social context in which interpretation occurs. If this view, central to the new philosophy of science, is accepted, there is no alternative but to regard the products of science as social constructions like all other cultural products. Accordingly, there seems every reason to explore how far and in what ways scientific knowledge is conditioned by its social milieu, how change of meaning is brought about and how knowledge is used as a cultural resource in various kinds of social interaction. It is, no doubt, possible that this revised view is fundamentally wrong and its conclusions merely a result of defective philosophical analysis. If this is so, the attempt to explore the social construction of scientific knowledge will probably founder. But not to make the attempt, at a time when the earlier set of philosophical ideas appears clearly to be inadequate, is an indefensible position.

In order to strengthen this conclusion, let me refer back briefly to part of the discussion in the previous chapter. There I described the crucial distinction made by many sociologists of knowledge between scientific and social/historical thought. The latter was regarded as amenable to sociological analysis because it had the following characteristics. All social thought, it was suggested, is related to a particular social context and undertaken from a particular, historically unique perspective. The knowledge gained is restricted by, and formulated in terms of, the observer's necessarily limited interpretative framework. Another way of putting this is that the answers obtained depend on the questions posed and on the questioner's presuppositions. Particular assertions can only be understood in the light of these background assumptions and any attempt to assess their validity must utilise specific rather than generally applicable criteria. Owing to the diversity and evolution of interpretative frameworks, meanings are 'made and ever re-made' in the course of social life and past contributions are continually reinterpreted. These, then, are the main characteristics attributed by sociologists to social constructions, that is, to mental products having their origin in the social as opposed to the physical world. They were thought to indicate that only socially variable knowledge of this kind was open to sociological analysis, because only here did participants' socially derived perspectives appear to penetrate into the meaning and evaluation of particular assertions. However, the similarity between this characterisation of social thought and the newer philosophical account of science is very obvious. If the revised view of science is accepted, the basis for the *traditional* distinction between scientific and social thought is eliminated, as is the exclusion of scientific knowledge from sociological interpretation (Barnes, 1974). Of course, we would hardly expect any other conclusion, for one of the central claims of the revised view is that scientific assertions are socially created and not directly given by the physical world as was previously supposed. (It is not implied that there are no differences at all between the analysis of the physical and the social worlds; only that at this level of generality the distinction traditionally used in the sociology of knowledge no longer seems to apply.)

The implications of this conclusion are considerable, for it enables us to ask a much wider range of questions than before about the social nature of science. We can ask, for example, to what extent presuppositions which are widespread in modern society have been implicit in scientific research and moulded its findings. We can examine exactly how scientists decide on the adequacy and significance of knowledge-claims, and whether their assessments are as disinterested as has been customarily supposed. We can study how the meaning of scientific assertions is reinterpreted in different social situations and whether, for instance, such reinterpretation can serve as a source of social power. In general, we can assess how far detailed sociological investigation of the social life of science provides support for the revised philosophical view; and we can explore what difference the revised view makes to our understanding of the social relationships involved in the creation of scientific knowledge. These are some of the issues which have recently begun to appear worth investigating. I will pursue them a little further in the chapters which follow, thereby taking a few tentative steps towards including science fully within the scope of the sociology of knowledge.

3

Cultural Interpretation in Science

In this chapter I will discuss some recent sociological work on the social and cognitive dimensions of the scientific research community. My aim is to sketch the outlines of an analysis of the social production of scientific knowledge which is based on detailed empirical evidence and which is consistent with the new philosophy of science. I will also try to show that the analysis which begins to emerge re-establishes links between the sociology of science and the sociology of knowledge. This theme will be taken further in the final chapter. I make no claim to provide a complete review of the relevant literature in these two chapters. Rather, I will select for close attention those contributions which seem to me to be particularly important in connection with my own line of argument.

In Chapter 1 I argued that when sociologists take for granted the standard view of science, they are thereby inclined towards a particular portrayal of the normative structure of the scientific research community. In the next section I will try to show that this classic portrayal of 'the scientific ethos' is inadequate and that an alternative interpretation can be formulated which is more consistent with the available evidence and which lends support to the account of scientific knowledge given in the last chapter. (Much of the discussion in this section is taken from Mulkay, 1976a.)

THE SOCIAL RHETORIC OF SCIENCE

As I tried to show in the opening chapter, the entire structure of moral and technical norms in science has been conceived as implementing what is taken to be the ultimate goal of science, namely, the establishment of objective knowledge of the physical world. Given the customary assumption that the scientific community has achieved knowledge which is closely and increasingly 'isomorphic to the structure of reality', it has been seen to follow necessarily that the members of that community have been predominantly open-minded, disinterested, impartial, independent, self-critical, and so on, in their intellectual endeavours. Assuming also that behaviour of this kind will not occur spontaneously throughout an intellectual community, it has seemed preferable to regard these attributes as characteristics of the community as such, that is, as norms which define the social expectations to which scientists are generally obliged to conform in the course of their professional activities. (Readers unfamiliar with these norms should consult Mitroff, 1974.)

In order to show that conformity to these norms is an essential feature of modern science, those presenting this argument tend to describe the negative consequences of deviant acts. The central idea is that actions which contravene such norms will clearly distort any resulting knowledge-claims. For instance, if scientists become too committed to their own ideas, that is, if they fail to abide by the norm of emotional neutrality, they will be unable to perceive when their ideas are inconsistent with reliable evidence. Similarly, if scientists adopt personal, non-universalistic criteria in assessing knowledge-claims, their judgements will be biased and will tend to diverge from the objective realities of the physical world. At the same time, if secrecy and intellectual theft were to exist to any extent in science and the norm of communality ceased to be an effective guide for social action, then it seems likely that the smooth and impartial extension of certified knowledge would be disrupted. As long as one remains within the epistemological framework of the standard view of science, it is not difficult to find reasons why departure from any of the normative principles listed above will tend to interfere with the creation of valid knowledge about the empirical world. Thus from this perspective the normative structure is the crucial feature of the scientific community. The norms of science are seen as prescribing that scientists should be detached, uncommitted, impersonal, self-critical and open-minded in their attempts to gather and interpret objective evidence about the natural world. It is assumed that considerable conformity to these norms is maintained; and the institutionalisation of these norms is seen as accounting for that rapid accumulation of reliable knowledge which has been the unique achievement of the modern scientific community.

The standard philosophical view of science favours the assumption that once certain major sources of distortion have been removed, it is fairly easy to recognise by means of systematic observation the empirical regularities of the external world. Accordingly, most of the normative principles postulated by sociologists have been conceived as minimising the impact of potential sources of distortion. This is one reason why the exposition of these norms tends to take the negative form outlined in the preceding paragraph; the necessity of conformity to the norms is demonstrated by showing that if scientists are *not* disinterested, humble, emotionally neutral, intellectually independent, etc., their perception and their judgement of reality will suffer. Thus the normative structure of science is seen as ensuring, as far as is humanly possible, that the external world is allowed 'to speak for itself'. The newer philosophy of science, however, provides much weaker grounds for inferring that scientists must abide by this set of norms. From the newer philosophical perspective the physical world is not so much revealed as socially and intellectually constructed. Take the principle of emotional neutrality, for example. Recent philosophical analysis stresses, not simply that complete neutrality is impossible, but that considerable commitment is necessary before even the simplest kind of observational work can begin. Moreover, as researchers probe their chosen phenomena in greater depth and detail, so their reliance on unexamined assumptions is likely to increase. At the same time, it follows from the newer philosophy of science that normative principles will acquire some of their meaning from the intellectual context in which they are implemented, that is, in part from members' specific scientific commitments. What is to count as emotional neutrality, impartiality or disinterestedness may well vary in line with scientists' research skills and interpretative frameworks. For example, scientist A may object that scientist B did not act in a disinterested fashion when he failed to recognise or cite A's work. But B may respond by claiming that he ignored that work, not out of any intent to gain an advantage over A, but because it was fundamentally misconceived and could only serve to confuse and mislead other researchers less well informed than himself.

To some unknown degree, therefore, the meaning to participants of these normative principles may depend on intellectual commitments and may be socially variable within science. This is likely to be particularly true of the principle of universalism, so central to the established sociological interpetation, according to which scientists are expected to judge knowledge-claims by means of impersonal, pre-established criteria. For if, as appears often to be the case, available criteria are unclear, or not easily applicable to particular instances, or different persons are using different criteria, it seems impossible for scientists to employ this principle in practice; even though, once reasonable consensus has been achieved, scientists may be able to formulate *ex post facto* the criteria which they have finally agreed are appropriate to a particular body of knowledge. In other words, the sociological notion of 'universalism' presupposes that technical criteria are generally available in science, in such a way that firm, impersonal judgements can be made with respect to most knowledge-claims and, thereby, with respect to the rewards and facilities which scientists deserve. The newer philosophy of science, however, by emphasising that the establishment of scientific knowledge is a creative process in which prior standards are frequently modified and new social meanings created, makes this presupposition questionable.

The general point that I wish to make is that the revised philosophical account of science, unlike the standard view, does not lead obviously and directly to the customary characterisation of the scientific ethos accepted for many years by sociologists. From the newer philosophical perspective there is no value in identifying a set of social norms which is designed to minimise distortion; partly because science is not conceived as an enterprise concerned with the definitive representation of reality, but also because it would be assumed that the meaning of normative principles would vary in accordance with changes in interpretative context. Accordingly, if we wished to formulate a set of scientific norms within the new framework of philosophical assumptions, these norms would undoubtedly be significantly different in form and content from those used in the past. I do not, however, intend to do this. For there are good grounds for revising much more radically the whole idea of normative regulation in science. By following this line of thought we will arrive at an account of the cultural resources of science which is not only consistent with the philosophical position outlined above, but which will also help to bring science firmly within the scope of the sociology of knowledge.

In recent years, there has been some criticism of the original analysis of the scientific ethos. One reason for such criticism is that detailed study by historians and sociologists has shown that in practice scientists deviate from some at least of these supposed norms with a frequency which is remarkable if we presume that the latter are firmly institutionalised. Another reason is that none of the empirical studies designed to discover how far samples of scientists express agreement with verbal formulations of the norms has produced evidence of any strong general commitment (Mulkay, 1969; but see also Storer, 1973). One response to findings such as these is to argue that the central normative element in science is furnished, not by this set of social norms, but by the scientific frameworks and technical procedures in terms of which the research community is internally differentiated. But this is not the only possible response. For we can argue that the original set of social norms was not so much wrong as incomplete. Merton, for example, has tried to account for the very considerable deviation from these norms by introducing the notion of 'counter-norm' (1973, ch. 18). Science he suggests, like other social institutions, does not employ a single set of compatible norms, but rather a series of conflicting pairs of norms. This response has been explored much more fully by Mitroff (1974), in the course of a detailed study of 'moon scientists'.

One of the important merits of Mitroff's study is that it provides a great deal of first-hand empirical material. In particular, it contains a large number of quotations from practising scientists. This means, not only that his own argument is exceptionally well documented, but also that it is possible to an unusual degree for the reader to extend Mitroff's own interpretation of his data. Mitroff shows, first of all, that the scientists in his sample do sometimes use variants of the norms described above, as standards for judging the actions of their fellows and as prescriptions about how researchers ought to behave. But the overwhelming import of his evidence is that, in addition, there exists in science an exactly opposite set of formulations and that conformity to these alternative formulations can also be interpreted, by participants as well as by observers, as being essential to the furtherance of science.

Let me give just a few examples.

Mitroff suggests that the norm of emotional neutrality is countered by a norm of emotional commitment. Thus many of the scientists studied by him said that strong, even 'unreasonable', commitment to one's ideas was necessary in science, because without it researchers would be unable to bring to fruition lengthy and laborious projects or to withstand the disappointments which inevitably attend the exploration of a recalcitrant empirical world. Similarly, the norm of universalism appears to be balanced by a norm of particularism. Scientists frequently regard it as perfectly acceptable to judge knowledge-claims on the basis of personal criteria. Instead of subjecting all research reports in their topic area to impersonal scrutiny, scientists regularly select out of the literature the findings of those colleagues whose work, for one reason or another, they have come to regard as reliable. In other words, scientists often regard it as proper to judge the man rather than the knowledge-claim. Once again, the counter-norm can be seen as functional, on the grounds that it saves researchers' time and effort, speeds up the rate at which research develops, yet at the same time ensures that greater weight is given in general to the judgement of those scientists who are perceived by their colleagues to be 'more able' or 'more experienced'.

Let me give one more example. Mitroff produces evidence to show that the ideal of common ownership of knowledge is balanced by a norm in favour of secrecy. He also suggests that secrecy, far from hindering the advance of science, actually contributes to this objective in several ways. In the first place, by keeping their results secret, researchers are able to avoid disruptive priority disputes. Secondly, attempts by others to steal or appropriate a scientist's work serve to confirm the significance of that work and to motivate him to continue his efforts. Thirdly, by keeping their findings from others, scientists are able to make sure that their results are reliable without jeopardising their own priority and, thereby, without undermining their enthusiasm for further research. Various supplementary arguments of this kind supporting each 'norm' are available to scientists as well as to the sociological analyst.

Mitroff's central argument, then, is that there is not one set of norms in science but at least two sets. The first set has been more or less accurately identified by those working in the functionalist tradition. But to describe the ethos of science in terms of this first set alone is to produce an account of science which is quite misleading; for each of the initial set of norms is matched by an opposing principle which justifies and prescribes actions in complete opposition. Thus common access to information is not an unrestricted ideal in science; it is balanced by rules in favour of secrecy. Intellectual detachment is often regarded as important by scientists; but no more so than strong commitment. Rational reflection is seen as essential; but so are irrationality and free-ranging imagination; and so on. In Mitroff's view, the empirical evidence which he has produced requires us to conceive of the scientific community as governed by these two major sets of norms and to interpret the dynamics of this community in terms of the complex interplay between these normative structures.

There can be little doubt that this evidence prevents us from accepting the initial set of norms alone as the normative structure of science (see also Blissett, 1972). However, I wish to argue that there are no compelling reasons for regarding either set of formulations or the two sets combined as providing the rules governing social life in science. This becomes clear as soon as we look more closely at the kind of evidence which is being adduced. Mitroff rightly criticises the procedure of extracting norms of science from the 'highly select writings of the rare, great scientists'. He suggests that we should derive 'the institutional norms of science' not only from the idealised attitudes of great scientists, but also from the messy behaviour and complicated attitudes found throughout the scientific community at large (1974, p. 15). He then proceeds to formulate his counter-norms by selecting out certain descriptive and prescriptive statements made by participants which appear to contradict the original set of norms. It is clear, therefore, that both sets of formulations are used by scientists to describe and to judge their own actions and those of their colleagues and to prescribe correct professional behaviour. But the mere use by participants of these verbal formulations does not demonstrate that they are the 'institutional norms' of science.

Social norms can be regarded as institutionalised when deviance is penalised and when conformity is regularly rewarded. Clearly the kind of analysis we have been considering so far assumes that norms and/or counter-norms are institutionalised in this sense; for otherwise it would be difficult to see them as making essential contributions to the extension of certified knowledge and to the progress of science. It is assumed, as Storer puts it, that there is an efficient 'internal police system' in science (1966, p. 85). However, when we examine the considerable literature on the allocation of professional rewards and the dynamics of social control in science, we find little indication that receipt of such rewards is in practice conditional on scientists' having conformed to the supposed norms or to the putative counter-norms in the course of their research.

The allocation of institutional rewards in science is closely associated with the system of formal communication. Scientists convey to their colleagues information which they believe to be interesting and reliable by means of the professional journals. Although there is also a considerable informal exchange of information, scientists are able to establish a conclusive claim to the credit for a particular contribution only by publishing it formally under their own names (Merton, 1973). In return for information which is judged to be of value, scientists receive professional recognition in various forms and are thereby able to build up a personal reputation, which can in turn be used to obtain other scarce resources such as students, research funds and academic promotion (Gilbert, 1977a). Perhaps the most important feature of this system, in the present context, is that the main medium of formal communication, the research paper, is written in a strict conventional style which concentrates attention on technical issues. Accordingly, references to the opinions, interests or character of the author are rigorously excluded. The report is typically written in the passive, so that allusions to the actions and choices of the author do not occur. The effect of such devices is to produce an aura of anonymity, so that the research becomes 'anyone's' research (Gilbert, 1976a).

There are, then, well-established norms governing the style of formal communication in science. But these must not be mistaken for norms which regulate the social dynamics of research in general. As Medawar has pointed out, the impersonal conventions of the research paper not only 'conceal but actively misrepresent' (1969, p. 169) the complex and diverse processes involved in the production and legitimation of scientific findings. This divergence between the formal procedures of communication and the actual social relationships involved in research exists partly because the rules governing the formulation of research reports make it virtually impossible for scientists to pass moral judgements, on the basis of published findings alone, about the author of a report. Thus their response to a published report and their allocation of recognition to its author, in the absence of other information about the author, cannot be influenced by the latter's conformity whilst carrying out his research to any particular set of social norms. In short, the conventional form of the scientific paper, which eliminates any reference to authors' conduct in carrying out their research, actually works to prevent scientists from allocating rewards in accordance with contributors' conformity to or departure from any particular code of ethics. Of course, as Mitroff demonstrates, scientists who are in informal contact do regularly make moral judgements of their fellows along at least two major dimensions. As we shall see later, these informal judgements may well affect the way in which scientists respond to others' results and may accordingly influence the allocation of rewards. However, in the course of informal inter-action, participants can choose freely from either set of norms. There is, therefore, no reason to expect that these informal processes will produce general conformity to either one of these opposing normative orders. Thus, it is not surprising to find that detailed empirical study has revealed no clear connection between conformity to any particular set of social norms and the receipt of professional rewards (Cole and Cole, 1973).

It appears, then, that conformity to most of the supposed norms and counter-norms of science is largely irrelevant to the institutional processes whereby rewards are distributed. Researchers are simply rewarded for communicating information which their colleagues deem to be useful in the pursuit of their own studies. There are no institutional mechanisms for rewarding in any direct way conformity to either set of social norms; nor is it possible to show that the provision of acceptable information *presupposes* the implementation of either set, for, as Mitroff demonstrates, both *contradictory* sets can be interpreted as being thus presupposed. But if the 'norms' and 'counter-norms' described in the sociological literature are not elements in an institutionalised normative structure, how are we to interpret the evidence presented by Merton, Mitroff and others? One answer to this question has already been suggested, that is, they are undoubtedly relatively standardised verbal formulations which are used by participants to describe the actions of scientists, to assess or evaluate such actions and to prescribe acceptable or permissible kinds of social action. But standardised formulations of an evaluative kind never govern social interaction in any straightforward fashion. This point has been made strongly by Gouldner:

... moral rules are not given automatic and mechanical conformity simply because they, in some sense, 'exist' ... conformity is not so much given as *negotiated* ... The rule thus serves as a vehicle *through which* ... tension is expressed ... there is usually more than one rule in a moral code that can be claimed to be relevant to a decision and in terms of which it may be legitimated. A central factor influencing one's choice of a specific rule to govern a decision is its expected consequences for the functional autonomy of the part ... What one conceives to be moral, tends to vary with one's interests. (1971, pp. 217-18)

The relevance to science of the general argument summarised in this quotation can be illustrated by referring once again to the discovery of pulsars. When the first paper on pulsars was published in 1968 by the radio astronomy group at Cambridge, there were numerous accusations of secrecy from members of groups in competition with the astronomers at Cambridge. It was said that the Cambridge group had unduly delayed publication; that they published insufficient data to allow other groups to undertake supplementary research; that they should have passed on their results before publication to close colleagues in neighbouring laboratories; that their secrecy prevented them from obtaining valuable advice from others; and that their action tended to impede the advance of science. Members of the Cambridge group, however, were able to provide various principles justifying their actions. In the first place, it was claimed that it was perfectly legitimate generally to avoid passing on information which could lead to anticipation by others. Secondly, secrecy was justified on the grounds that it gave researchers time to check their results and to publish high-quality work, thereby ensuring the smooth development of scientific knowledge. Thirdly, it was said to be legitimate to make sure that important results improved the reputation of one's own group and its ability to obtain research funds. Fourthly, it was also said that scientists had the right to protect the first achievement of a young research student or the right of observers to have the first attempt at interpreting their own findings. Fifthly, it was said that steps had to be taken, in the particular case of pulsars, to prevent the press from misrepresenting this remarkable discovery. As we would expect in view of this confusing variety of diffuse and overlapping rules, some participants denied that there had been any undue delay in making public the first observations of pulsars.

In the original sociological study from which this material is taken, the conclusions about the operating principles relevant to the communication of research results are summarised as follows.

> There does not appear, however, to be a general commitment to these principles; nor are there clear procedural rules governing the communication of results. As a result, misunderstanding and resentment are sometimes produced by what are variously regarded as secrecy or as legitimate control over the circulation of scientific information. (Edge and Mulkay, 1976, p. 250)

It is worth noting in connection with the argument above that conformity to social norms is irrelevant to the receipt of rewards, that despite the heated differences of opinion at the time of the pulsar discovery about the propriety of the actions of the Cambridge group, six years later two of its members received a Nobel Prize based in large part on this discovery.

In science, then, we have a complex moral language which appears to focus on certain recurrent themes or issues; for instance, on procedures of communication, the place of rationality, the importance of impartiality and of commitment, and so on. But if the example described immediately above is representative and if the preceding argument is correct, then no particular solutions to the problems raised by these issues for participants are firmly institutionalised. Instead, the standardised verbal formulations to be found in the scientific community provide a repertoire or vocabulary which scientists can use flexibly to categorise professional actions differently in various social contexts. A major influence upon scientists' choice of one verbal formulation rather than another, as Gouldner notes, is likely to be their interests or objectives. It can be assumed that, for a given scientist or group of scientists, these interests will vary from one social context to another. Thus, in the example given above, when researchers were frustrated by the apparent reluctance of others to make significant findings available to them, they tended to select principles favouring communality which justified their condemnation of the others' behaviour and added weight to their own exhortations. In contrast, those scientists who had made the discovery were able to find principles in favour of personal ownership of results. In different circumstances, a person's or a group's choice of rules can be entirely reversed. Not only is it possible to vary one's choice of formulations as one attempts to identify the evaluative characteristics of different acts, but it is also possible to apply different formulations to the same act as one's social context changes. For instance, it is possible to 'justify' keeping certain results secret on the grounds that to do otherwise would jeopardise a graduate student's first efforts; and then subsequently, when recognition or prizes are being allocated, to maintain that in *this* context the student's contribution should be regarded as unimportant.

It must not be imagined for one moment that I am in any way accusing scientists of bad faith or of being less moral than other social groupings. I am simply arguing that within the relatively distinct community concerned with scientific research, as indeed in most areas of social life, interaction cannot be adequately depicted as expressing any one or more sets of institutionalised normative principles or operative rules deriving from such principles. It seems more appropriate to portray the 'norms of science', not as defining clear social obligations to which scientists generally conform, but as flexible vocabularies employed by participants in their attempts to negotiate suitable meanings for their own and others' acts in various social contexts. The details of these social dynamics in science are not yet well understood. For instance, it may be that the ability to control others' access to valued information may increase scientists' capacity to gain acceptance for their preferred categorisations. This may have been true to some extent in the case of pulsars. But until we move away from the traditional account of the scientific ethos it is hardly possible to conceive that such exercise of power plays any significant part in the scientific community. There has, therefore, been little investigation of phenomena of this kind. Nevertheless, although analysis of the negotiation of social categories in science is still in its infancy (Law, 1976), what *is* clear is that it is highly misleading to regard the diffuse repertoire of standardised verbal formulations as the normative structure of science or to maintain that it contributes in any direct way to the advance of scientific knowledge.

This last point returns us to the discussion with which this section began. In abandoning the customary characterisation of the scientific ethos, we have removed any apparent inconsistency between the sociological analysis of scientific norms and the philosophical analysis of scientific knowledge. But this leaves an appreciable gap in the sociological interpretation of science. For the traditional conception of the scientific ethos, although it avoided any close examination of scientific knowledge itself, did at least provide some kind of general account of the latter's genesis: namely, as the necessary by-product of widespread conformity to the supposed norms of science. Clearly, then, as this view of the social origins of scientific knowledge must be rejected along with the traditional idea of the scientific ethos, an alternative sociological analysis of the production of scientific knowledge is needed. In the next section, I will examine a number of recent case studies of scientific development which draw attention to features quite unlike those assumed in orthodox sociological analysis and much closer to the conclusions of the new philosophy of science. I have no intention, however, of attempting to provide a complete, formal analysis of the social production of scientific knowledge.

THE DYNAMICS OF KNOWLEDGE-PRODUCTION

When we examine the overall growth of modern science one thing stands out clearly, namely, that there has been a continuous creation of new areas of investigation and new realms of knowledge (Price, 1963). In the seventeenth century, the entire physical world fell within the scope of 'natural philosophy'; and one man could encompass in his studies the full range of available knowledge. By the end of the nineteenth century this had become quite impossible. Knowledge of the natural world had become much more extensive as well as more detailed and complex (Mason, 1962); and the major scientific disciplines had crystallised into more or less distinct intellectual domains, each of which was separately established in the centres of higher learning with control over professional training and over access to its own area of knowledge. This process of intellectual and social differentiation has continued up to the present day; so that now each discipline is further sub-divided into many specialties. Each of these specialties is composed, in turn, of numerous specific areas of research, most of which deal with phenomena unknown a generation before. It is true that some of the most well-known advances in scientific thought have not involved differentiation so much as the reconceptualisation of existing bodies of knowledge (Kuhn, 1962). Nevertheless, to a very considerable extent, scientific knowledge has developed by the identification and detailed investigation of phenomena which have not been known to exist previously or which have not been studied before in any depth. The *typical* pattern of growth, then, in science is not the revolutionary overthrow of an entrenched orthodoxy, but the creation and exploration of a new area of ignorance (Holton, 1973, ch. 12). Within many such areas there occurs a gradual movement through three discernible, although overlapping, stages; that is, from an initial phase of exploration, through a stage of unification and into a final period of decline (Mulkay, Gilbert and Woolgar, 1975). Although this sequence is by no means inevitable (Law, 1976), it is characteristic of most areas where consensus, and therefore certified knowledge, is actually achieved.

If we are to understand how scientific knowledge is socially produced, there are advantages in concentrating particularly on the early stages in the recurrent movement towards intellectual consensus in science. Collins elucidates this point with the help of an analogy.

> When we consider the grounds of knowledge, we do it within an environment filled with objects of knowledge which are already established. To speak figuratively, it is as though epistemologists are concerned with the characteristics of ships (knowledge) in bottles (validity) while living in a world where all ships are already in bottles with the glue dried and the strings cut. A ship within a bottle is a natural object in this world, and because there is no way to reverse the process, it is not easy to accept that the ship was ever just a bundle of sticks. Most perceptions of the grounds of knowledge are structured in ways derived from this perspective. (1975, p. 205)

Collins goes on to argue that in the study of science we can to some extent escape from this predicament, because we can identify fairly easily those scientists working on a common problem and we can therefore examine contemporary scientific developments as they proceed. I intend now to look at some of Collins's empirical material, as well as other recent sociological work, in order to see what is revealed by close study of the social production of scientific knowledge. It will become evident that the conclusions reached in these studies help to extend the argument developed in the preceding section.

Collins's first study (1974) is of a social network of scientists concerned with building a special kind of laser called a 'TEA laser'. This network was exploring a new area in the sense that, although various kinds of lasers had been built before so that general notions were available of what lasers in general ought to do, this particular form of laser had only recently been made operational. In 1970 an effective TEA laser was reported in the literature for the first time and numerous groups began to try to produce their own versions. At the time of the study some had succeeded and some had not. Collins concentrates on the transmission of information among these groups. His central finding is that even those scientists who had actually produced a working laser were unable to make fully explicit the knowledge which was required. In many cases, of course, scientists were simply *unwilling* to formulate their technical knowledge openly. They preferred not to pass on all that they knew, because to have done so would have reduced their competitive advantage. But there were also cases where participants whose lasers were operative seemed to be *unable* to convey their knowledge to others. Thus one scientist would help another to build a machine which seemed to both parties to be identical with his own successful laser, only to find that inexplicably it failed to work. Furthermore, no group managed to produce an operating laser on the basis of formal published information alone. Success always depended on direct personal contact. These contacts often had to be repeated many times until, unpredictably, the laser could be deemed to be operating properly (Collins and Harrison, 1975). Collins suggests that personal contact was essential because only through direct interaction could scientists communicate the tacit and unformalised knowledge on which their work depended.

It seems, therefore, that in this area at this early stage it was impossible for participants to know whether a scientist had absorbed the required knowledge except through his successful interaction with other specialists. Members' scientific competence and compliance with technical standards could not be judged fully by the application of pre-established, formal criteria; but only by *ex post facto* negotiation with those other participants who were already regarded as competent judges. An interesting question here is whether this network's reliance on informal processes and tacit knowledge would have been less visible, without necessarily being less important, at a later stage when all interested groups had an operative laser, and what looked like a set of 'principles of construction' had been ostensibly agreed. It may well be that at this later stage both participants and analyst would have been much more likely to portray the construction of TEA lasers as due to the implementation of an unambiguous series of specifiable instructions. But this is speculation. Collins emphasises only that it seems inappropriate to describe participants at the time of his study in the traditional sociological manner as carriers of independent, impersonal knowledge. 'The point is that the unit of knowledge cannot be abstracted from the "carrier". The scientist, his culture and skill are an integral part of what is known' (1974, p. 183). The tacit and personal character of much scientific knowledge and the social negotiation of what is to count as valid knowledge are examined further in Collins's study of research into gravitational waves (1975).

It is only recently that the phenomenon of gravitational waves has received more than passing scientific attention. Up to about 1969 only one scientist had tried to detect this form of cosmic radiation. However, after this man published a formal claim to have observed gravitational waves, numerous other scientists quickly entered the field and devised their own observational equipment. There was fairly complete agreement among participants in this field on certain general points. Most of them agreed, for example, that gravitational waves are predicted by Einstein's general theory and that certain catastrophic astrophysical events should release such waves. There was also broad agreement about the kind of observational apparatus required and about the classical physics which underlies these experimental tools. In addition, all participants appeared to endorse similar general criteria of validity; for example, they agreed on the need for controlled experiment and for theoretical interpretation which was consistent with the experimental evidence. This seems, then, to have been normal science, that is, an attempt to fill in the details of a well-established paradigm by means of conventional observational methods. Yet, Collins stresses, the work involved was in no way routine, for most participants were fundamentally uncertain about how the existence of gravitational waves could be demonstrated and the network was deeply divided about the meaning of its members' experimental results. Thus neither established knowledge nor formal criteria provided any unambiguous means of distinguishing valid from invalid claims. Both these kinds of cultural resources were interpreted in diverse ways when applied to the new case of gravitational waves.

Collins is particularly concerned to counter what might seem to be the obvious interpretation of the activities of scientists in this field, namely, that they were engaged in straightforward replication or refutation of the original claimant's results. The difficulty with this interpretation is that, in practice, replication or disproof can only be achieved when there is considerable agreement about the meaning of observations, the adequacy of experimental procedures, and so on. For replication to occur, participants must be able to decide what are to count as equivalent and reliable observations. Much of the research on TEA lasers is fairly close to the usual conception of replication because, in this case, there was a criterion against which success or failure could be measured and there were 'competent judges' with whom agreed definitions of 'success' and 'failure' could be reached; although even here the criterion was difficult to specify and its application was a matter for negotiation. But neither agreed criteria nor competent judges were available in the case of gravitational waves.

Collins produces a variety of evidence to demonstrate that there was no common assessment of the experimental procedures or results of any of the members of this network. What one scientist viewed as interesting, another dismissed as unimpressive and a third violently rejected as outright fraud. Accordingly, nobody was at all interested in attempting to repeat in detail the original experiment which had provided the impetus for further research in the area. There was no point either in duplicating or in failing to obtain a result which had no scientific meaning. Thus participants were more directly concerned with devising some new kind of observation which they thought had a better chance of being recognised as a 'competent measure of gravitational waves'.

Although most participants clearly thought of themselves as engaged in checking in some way the original knowledge-claim as well as subsequent findings on gravitational waves, Collins suggests that they did this indirectly by entering into 'negotiations about the meaning of a competent experiment' (1975, p. 216). Because there was such a variety of scientific opinion in the field, members' primary concern was to establish what should count as a 'working gravity wave detector'. If they were to succeed in doing this, Collins points out, they would provide an effective interpretation of the phenomenon of gravity waves. This can be illustrated by the simple example of experiments on heat. Because there is general agreement about the nature of heat, a great variety of superficially different activities are seen by scientists as being equivalent in that they are all 'competent temperature measuring experiments'. In treating different measures as equivalent, for example, dipping a glass tube filled with mercury *or* two metals linked by a voltameter into a liquid, scientists are drawing on their knowledge about the characteristics of heat; and the complete set of equivalent activities goes some way towards defining the nature of heat. Furthermore, if some experiments and measures which are now excluded from the class of 'competent experiments' were to be included, it would follow that the accepted properties of heat would have changed (Collins, 1975, p. 217). In the case of temperature measurement, and to a lesser extent laser building, the technical culture is sufficiently well established for participants to distinguish competent from incompetent work and valid measures from invalid measures with (comparative) ease. But this was not so in the case of research on gravity waves. Thus as researchers in this latter field negotiated agreement about which experiments were to be regarded as competent and equivalent, they were defining the nature of their problematic empirical phenomenon and creating a distinctive region of scientific culture.

Collins pays little attention to clarifying in detail the form that negotiation takes in this field. But the picture seems to be roughly as follows. Participants bring to the study of gravitational waves certain general assumptions which they appear to share. As we have seen, these include such varied interpretative resources as Einsteinian theory and belief in experimental methods. These resources are brought to bear on the largely unexplored phenomenon of gravitational waves; but they yield no unique or unambiguous interpretation. Participants do have enough in common to concentrate on a limited range of empirical variables which they are able to see as related, as being members of the same set. But Collins shows that these variables are employed selectively and interpreted differently in arguments favouring different knowledge-claims. What one participant takes as given, another takes as problematic and even as undermining the adequacy of the other's claim.

Thus each experimenter (or research group) uses the available technical culture and his own expertise in a flexible manner to reveal the inadequacies of others' findings and to support his own claims. Furthermore, those involved seem not to distinguish clearly between technical and non-technical criteria of evaluation (see also Bourdieu, 1975). For judgements about the merit of knowledge-claims are said by participants to depend also on personal considerations such as one's faith in an experimenter's capabilities and honesty, one's views about his personality and intelligence, his reputation, social location and psychological approach, one's access to 'inside information', and so on.

Of course, as we saw in the previous section, it is unlikely that these verbal formulations determine scientists' technical assessments in any strong sense. It is more probable that they are a way of strengthening the presentation of an evaluation already conceived. However, the same point seems to apply equally to the use of scientific and technical formulations (see also Bloor, 1976). Both the social and the technical culture of science appear to provide members with flexible symbolic resources which can be, and are, combined to devise a considerable variety of interpretative positions in connection with a common research problem. Scientists do have a social and technical culture which provides something like a shared context in this area of research. Thus certain *unorthodox* notions, such as the existence of a 'fifth physical force' or the operation of psychic forces, although sometimes used informally to explain particularly difficult results, were seldom allowed into the public forum. They were not part of the accepted repertoire. But the precise meaning of the orthodox cultural repertoire has to be established anew in this emergent field by processes of symbolic interpretation and negotiation. As Collins puts it, quoting from McHugh: any consensus which ensues 'is conceivable only as a socially organised upshot of contingent courses of linguistic, conceptual and social behaviour' (1971, p. 329).

One of the important merits of Collins's work is its use of contemporary data. He provides us with an unusual view of science by recording what happens informally, before a firm scientific interpretation has been accepted and before an established body of knowledge has come to appear to be the only rational possibility (see also Kemp, 1977). Several other recent studies, despite being retrospective rather than contemporary, have reached similar conclusions. Such retrospective studies have one major advantage over Collins's work, namely, they can more easily follow the process of negotiation through its full sequence. They can, therefore, examine how consensus is achieved, maintained and sometimes abandoned. Gilbert's study (1976b) of radar meteor research, for instance, shows how consensus emerged with respect to a major issue (the origin of sporadic meteors), even though central participants both for and against this consensus seemed at the time to regard the available scientific evidence as incomplete or inconclusive. In other words, the attainment of consensus in this case was not due solely to intellectual considerations. In another retrospective study, Pinch (1976) looks back at quantum mechanics in the 1950s and is able to show how physicists were able to defend their established framework of interpretation against a threatening knowledge-claim by a highly selective employment of theoretical resources in different social contexts. This case, he suggests, supports 'the view that scientific theories themselves are multi-dimensional and that what constitutes a theory in science is a variable and will mean different things to different groups of scientists'.

A particularly well-documented study of this kind, which I will now examine in some detail, is that by Wynne (1976) of Barkla and the J phenomenon. Wynne's objective is to revise the prevailing account of 'the J phenomenon affair' and to correct the interpretation current in the 'oral folklore' of the scientific community. This latter interpretation depicts the dispute about the J phenomenon as being resolved in favour of Barkla's opponents by the application of impersonal rules, which are independent of social context and which distinguish unequivocally between truth and error. Wynne argues, in contrast, that the 'scientific reasons' given in the literature for rejecting Barkla's claims are better conceived as a rhetoric seeking to justify a rejection based on other considerations. These other considerations, he suggests, include social factors and point towards 'a different version of scientific rationality and knowledge' (1976, p. 308).

There are two distinct phases in the history of the J phenomenon. In the first phase, between 1912 and 1923, the eminent physicist Barkla proposed a theory to explain a new set of X-ray emissions, the 'J radiations', emanating from the electrons of a specific 'shell' or series in the atom. This theory was formulated in terms of the 'classical' interpretation of X-ray scattering and initially it was widely accepted. By the early 1920s, however, observational anomalies had begun to appear. In particular, physicists using the increasingly popular spectrometer, which had been widely adopted because of its precision and the vast amount of experimental work which it opened up, failed to confirm Barkla's results. This was actually to be expected from Barkla's point of view; for the beams which he wished to measure were of such low intensity that, he reasoned, they would not be detectable by means of these new observational techniques. Mainly for this reason, Barkla had taken the unorthodox decision to use relatively 'old-fashioned' absorption methods which, although they were difficult to manipulate experimentally, had the crucial advantage of being more sensitive to low-intensity emissions. Wynne maintains that Barkla's unusual, but reasoned technical strategy was never taken fully into consideration in assessments of his work. Instead physicists simply used their own (inappropriate) 'technical norms' in an unself-critical manner to question the competence of Barkla's research and the validity of his knowledge-claims.

Before long, theoretical objections were added to these experimental difficulties. For in 1922 Compton proposed a new theory which, supported by unexpected empirical evidence and some precise and sophisticated calculations, shattered the classical theory of scattering. One major implication was that, if Compton's theory were accepted, there was no room theoretically for a J series in the atom. As Compton's interpretation came to attract widespread support and as opposition to the J series grew, Barkla undertook more absorption experiments and in 1923 he abandoned the J series *theory* as a valid explanation of his results. But he did not adopt Compton's theoretical analysis. Barkla continued to believe that there was an observational phenomenon, the J phenomenon, which still required proper theoretical interpretation.

Barkla's repudiation of his J series theory was not an entirely negative move, for he had seen 'the vague but tantalizing outline of things much more revolutionary' (1976, p. 314). Accordingly, over the next decade, he sought to develop an alternative theoretical model which differed from his own initial theory and from the dominant interpretative framework. During this second phase his work departed more and more from the research tradition of mainstream physics. He continued to reject the use of the spectrometer and some of the scientific assumptions associated with that technique. He maintained that the fashionable research technology forced nature to fit preconceptions built into the design of the major instrument and that the discovery of fundamental results was sacrificed for highly precise routine measurements. He stressed that heterogeneous X-ray beams did not always behave as a simple linear sum of individual wavelength components (a basic premise of orthodox spectrometer analysis), but rather as an 'organic whole'. In accordance with this concern for a more holistic approach he employed a scientific idiom derived in large measure from nineteenth-century *natur-philosophie*; and he continued to use an out-of-date and unpopular hydrodynamic metaphor, which expressed his belief in the continuity of nature but which was unsuitable for a physics community increasingly committed to quantum theory and the view that radiation is always emitted in discrete quantities.

During this phase of research into the J *phenomenon*, a considerable number of papers was published by Barkla and his students. But this work had no impact on the development of orthodox physics, except that several reviews were produced in which Barkla's claims were finally discredited. It is usually held that in these reviews Barkla's work is either shown to be explicable in terms of Compton's by then orthodox theory or to be so unsystematic and irregular as to call for straightforward rejection as incompetent research. The central point of Wynne's analysis is that these reviews do nothing of the kind and that, in fact, there is no explicit refutation available of Barkla's work which can plausibly be described as an impartial evaluation based on established standards of scientific rationality.

In the first place the most influential review, written by Dunbar, appeared in 1928, some five years before Barkla finished publishing results. It is, therefore, difficult to accept the popular opinion that Dunbar finally buried the J phenomenon. Secondly, all the reviews refer to Barkla's mistake over the J *theory* as if this were in some way evidence of his being mistaken again about the J *phenomenon*. Wynne points out that a 'strong claim is thus being made here against Barkla's credibility, a claim that borrows little from any of the usual prescriptions for the rational conduct of scientific disputes' (1976, p. 327). Barkla is also criticised for being unable to identify the causal factor which triggers off the organic mode of action of heterogeneous X-rays as well as for publishing incomplete accounts of his experimental arrangements. These criticisms are used in a highly selective manner. For, as we saw in the case of laser research, it is literally impossible to provide a complete description of experimental procedures and this is not usually required of experimentalists. Nor are observers always expected to give a causal explanation of their results before the latter are accepted as competent.

In addition to these relatively general charges, most reviewers provide a more detailed analysis of some of Barkla's empirical claims. But their conclusions at this substantive level are no more convincing. For instance Dunbar, although he noted that Barkla was attempting to treat complex radiations as organic wholes by means of appropriate techniques, used homogeneous beams for his 'replications' and thereby excluded the factor of heterogeneity which alone gave meaning to Barkla's experiments. For Barkla, Dunbar's work was not a replication of his own and had no bearing on the validity of his findings. For Dunbar and for most orthodox physicists, however, it was accepted as a clear refutation. Wynne also shows how Dunbar, in order to achieve his 'refutation' of Barkla's results, was forced to reinterpret and repudiate some of his own earlier work which he had previously presented as being quite unequivocal and free from the possibility of error. 'In marginal cases such as this, one sees what one is predisposed to see, and intepretational caution is cast to the winds in the tacit strategy of persuasion. But it is remarkable how the meaning of the old observations was changed diametrically, so as to support the new conclusions' (1976, p. 328). The logic of this kind of argument seems to be that of interpreting both empirical results and general standards of adequacy in accordance with a predetermined conclusion, in this case the falsity of Barkla's claims.

There are numerous other examples given by Wynne of Barkla's critics selectively interpreting data in a way that discredited their opponents' views. In addition, he shows how these 'empirical refutations' were given added force by being embedded in a web of critical comment which systematically undermined Barkla's scientific credibility. For example, Barkla was accused of wanting dogmatically to retain the classical theory of X-ray scattering and of being unwilling to give up his own J theory. Wynne shows that both these accusations were unjustified. One reviewer even used a mistake made by two of Barkla's research students to condemn his entire research school, suggesting that this erroneous paper was Barkla's only major support. The reviewer perhaps did not notice and certainly did not mention that Barkla never tried to use the paper in support of the J phenomenon. It is clear, Wynne contends, that the account usually given of the rejection of Barkla's work on the J phenomenon is quite misleading. Barkla's claims were never conclusively disproved by clear demonstration of their failure to meet impersonal standards of validity.

> The published refutations of Barka's work are marked by confusion as to the nature of his exact position (even to the extent that views set down in major journals were apparently not known); by standards of evidence which can only count as 'standards' to those already predisposed to accept their conclusions; by heavily implied sanctions against a scientific adversary which entail norms of a strictly non-rational kind; and by extravagant inconsistency as to why the phenomenon should be rejected. These published refutations should, I propose, be treated more as symptoms, as rhetorical rationalizations of a rejection already sealed for less tangible reasons. (Wynne, 1976, p. 335)

Having reached these conclusions with respect to the case of the J phenomenon, Wynne goes on to offer a somewhat speculative, but none the less interesting, general analysis along the following lines. Scientific consensus, and thus scientific knowledge, is not achieved by means of conclusive proof and disproof. Scientists are always faced with ultimate uncertainty and ambiguity. But intellectual commitments are necessary and are regularly made. They are not achieved, however, by the application of any set of pre-established, formal criteria. Adoption and rejection of research programmes is a much more pragmatic process and is greatly influenced by scientists' relatively local interests. For example, the spectrometer became a dominant research tool largely because of its practical advantages; it could be used routinely to provide precise measurements over a wide range of areas. Yet, once firmly established, it became a major resource for rejecting knowledge-claims which embodied different scientific assumptions. This means that 'orthodoxy lurches along via apparently arbitrary factors located in social (including technical) practices; and that traditional predispositions, socialized values, assumptions and practices, and the like, play a crucial part in the eventual route which "true" science is deemed to have taken. That route is only *formally* mapped out in hindsight' (1976, p. 336).

Wynne is here suggesting that the formal rationality of scientific knowledge, which requires that such knowledge be shown to conform to invariant principles of validity, tends to be brought to completion after the event. One important aspect of this operation is the final repudiation of any significant opposing views. In this way, the *appearance* of formal rationality is strengthened. This, in turn, has two further consequences. Within the research community, it fosters the concentration of intellectual effort. In addition, it helps the scientific profession to maintain its credibility in the wider society and to obtain economic and social support.

This last statement draws attention to the fact that the studies with which we have been concerned so far deal only with those processes occurring within relatively small and highly specialised research networks. Clearly these social networks are parts of a broader pattern and in the final chapter I will examine some of the links between scientific research and the wider society. For the moment, however, I shall extend the scope of the discussion only slightly, by looking at two studies in which some attention is given to the connections between specific specialties and the rest of the research community. The first of these studies is that by Collins and Pinch of parapsychology (1978). These authors stress that the *reality* of such paranormal phenomena as extrasensory perception or psychokinesis is irrelevant to their analysis. They are concerned solely with studying how scientists attempt to establish or refute knowledge-claims dealing with that class of phenomena referred to as 'paranormal'. In particular they argue that, by exploring this case, we will be able to see how scientific knowledge is the contingent outcome of both social and cognitive processes in the research community as a whole.

Claims by a recognised scientist to have demonstrated the existence of paranormal phenomena were first advanced in the 1930s. But general agreement about the meaning of basic observations or the scientific legitimacy of this kind of inquiry has still not been achieved today. One reason for this is that the debate, or negotiation, about paranormal phenomena has not been restricted to those doing detailed research, that is, to parapsychologists themselves. Many scientists working primarily in other areas seem to have regarded the claims of parapsychologists as a threat to established knowledge and to the credibility of science. Thus one pronounced feature of the debate has been the contest between parapsychologists, whose minimal claim has been that there are genuine phenomena to be scientifically investigated, and orthodox scientists who have sought to undermine the parapsychologists' entire enterprise.

Collins and Pinch distinguish two 'forums' in which this debate between orthodox and deviant scientists has been carried out.

On the one hand there is what we will call the 'constitutive' forum, which comprises scientific theorising and experiment and corresponding publication and criticism in the learned journals and, perhaps, in the formal conference setting. On the other hand, there is the forum in which are set those actions which—according to old-fashioned philosophic orthodoxy—are not supposed to affect the constitution of 'objective' knowledge. We will call this the 'contingent' forum, and would expect to find there the content of popular and semi-popular journals, discussion and gossip, fund raising and publicity seeking, the setting up and joining of professional organisations, the corralling of student followers, and everything that scientists do in connection with their work, but which is not found in the constitutive forum. (1978)

Within the framework of what I have called the standard view of science, which Collins and Pinch refer to as 'old-fashioned philosophic ortho-doxy', we would expect the success or failure of parapsychology to be determined solely in the constitutive forum. Parapsychologists would present the results of a number of carefully conducted experiments through the formal channels of communication and these results would be judged by the application of impartial canons of proof, consistency, adequate evidence, and so on. Presumably within a relatively short period of time, it would be possible for virtually all those involved, whatever their initial inclination, to decide objectively whether or not there were genuine phenomena to be investigated. Collins and Pinch show, however, that the debate has not been remotely like this. They go on to argue that the issue was not resolvable within the formal limits of constitutive debate and that both sides have in fact continually employed resources from the contingent forum in their attempts to establish as authoritative their opposing definitions of paranormal phenomena.

Let us look first at the critics of parapsychology. In the first place, it is clear that many of them preferred to keep the knowledge-claims of parapsychologists from being considered at all in the constitutive forum. For example, papers from parapsychologists were regularly rejected, not only when referees were evenly divided in their recommendations, but even on those few occasions when there was a majority in favour of publication. Moreover, when positive results were actually published, journal editors were inclined to indicate in one way or another that the journal was not endorsing these findings. But, as we have noted, the screening of papers by referees and editors is merely the first step in the process of evaluation. Collins and Pinch stress that formal presentation in itself need never lead to acceptance by determined opponents for, as we saw in the case of gravitational waves, evaluative resources are highly flexible and can be employed to support quite divergent scientific positions. Thus some participants simply saw the empirical findings as uninteresting. The so-called results were 'empty correlations' which were not worthy of further attention. For others, however, the decisive factor in rejecting the results of parapsychological research was that the empirical results which had been obtained had been given no theoretical analysis which was at all convincing. This was contested by many parapsychologists, most of whom saw themselves as working within an appropriate theoretical framework.

So far, then, we have noted clear differences of opinion about what constitutes an empirical finding in this area and what counts as a theoretical interpretation. In addition, it was possible, without doubting the validity of some of the experimental work by the accepted standards, to question the validity of the standards themselves. Thus one mathematician reasoned roughly as follows: 'Experimental results interpreted by classical probability theory appear to indicate the existence of paranormal phenomena. Although the experiments seem to be satisfactory, the conclusion cannot be accepted. Therefore, there is something wrong with probability theory'. If this line of reasoning were to be generally accepted it would imply that established procedures of inference over a wide area of science, namely, those fields which depend on probability theory, would have to be revised and the boundaries of certified knowledge extensively re-drawn. Few scientists, however, thought that there was any need to change techniques of statistical inference in response to findings which they saw as being blatantly false. A much more usual procedure was merely to assert that the existence of paranormal phenomena was inconsistent with incontestable scientific knowledge and, therefore, that any positive results, no matter how plausibly presented, must be the product either of experimental error, fraud or self-deception.

At this point, Collins and Pinch suggest, we can see the critics of parapsychology beginning clearly to deploy the repertoire of the contingent forum in the course of formal knowledge-constitutive debate. Thus the case of parapsychology is especially informative in sociological terms because, as a result of the new specialty's deviant and threatening appearance in the view of the orthodox, the merging of the contingent with the constitutive forum in the process of cognitive negotiation becomes unusually evident. For instance, critics have frequently asserted that parapsychology is merely an irrational cult maintained by faith, in contrast with genuine science which is based on evidence and demonstration. Parapsychology is said to be merely another form of spiritualism and belief in the occult. This characterisation enables critics to account conveniently for the fact that some physicists, psychologists, and so on, appear to have a different view of the available evidence and the conclusions to which it leads. Belief in the rationality and consensus of science is made consistent with apparently fundamental intellectual divergence, by viewing those with opposing ideas as, in the last resort, simply irrational.

Once advocates of parapsychology have been defined in this way, there is no need to treat its knowledge-claims seriously or to convince its practitioners of their errors. A similar kind of argument is that which accuses, not just individual parapsychologists, but the whole specialty, of fraud. From this perspective the critic may admit that the evidence occasionally *looks* compelling. But consistency is maintained between this evidence and what is taken to be established knowledge by asserting that the former is not and cannot be *real* evidence. As Collins and Pinch show, there is no defence against this argument when it is pushed to extremes. Critics can always show that results *could* have been fabricated, either in the course of normal interaction between the experimenter and his subject or between the subject and outside helpers, for example, by means of hidden transmitters.

> This tradition of giving credibility and persuasiveness to the fraud hypothesis by the demonstration of its *possibility* has become a standard procedure in the critique of parapsychological experiments ... The logic of the fraud hypothesis not only appears to remove any need for empirical tests from the scientific decision-making process but can also be put forward without any empirical evidence that fraud actually took place. (Collins and Pinch, 1978)

The tactics of the parapsychologists appear to have been broadly similar to those of their opponents. It seems that some parapsychologists, at least early in their research, had assumed that the existence of paranormal phenomena could be established experimentally, that is, by the formal publication of a definitive experiment. But as we have seen, any experimental results in this field can be rejected with the help of the interpretative resources available to orthodox scientists. And parapsychologists, as well as some of their opponents, have come to recognise this. Accordingly, parapsychologists have employed a variety of supplementary methods for establishing the legitimacy of their claims, from the relatively technical to the obviously social. At the technical level, they have increasingly adopted the complex experimental methods of physicists, biologists and psychologists. They have also refined their statistical procedures so that, as early as 1937, they were able to gain official endorsement by the Institute of Mathematical Statistics. Moreover, one of the central foci of interest among parapsychologists has been the attempt to reconcile their findings with orthodox electromagnetic and quantum-mechanical theory. These efforts within the constitutive context have been accompanied by action in the contingent forum. Collins and Pinch show that by the latter action parapsychologists have tried to influence the reception given to their knowledge-claims; they also suggest that contingent action has to some extent helped parapsychologists to establish telepathy, clairvoyance and psychokinesis as genuine scientific phenomena.

Collins and Pinch provide clear evidence that parapsychologists have sought systematically to obtain university posts in their subject, to recruit students, to acquire funds from legitimate sources, to publish in recognised journals, and to join prestigious scientific associations such as the AAAS. In addition, it appears that, as parapsychologists have increasingly obtained these social objectives over the years, the legitimacy of their intellectual endeavours has become more widely accepted. This has been particularly true in the USA, where the social organisation of science provides a relatively favourable environment for the establishment of new scientific groupings (Ben-David, 1968). Of course, the mere existence of a relationship between increasing scientific legitimacy and organisational expansion tells us very little. Collins and Pinch claim that parapsychologists have gained increasing intellectual acceptance by obtaining most of the *social* characteristics of other legitimate scientific specialties. But clearly the process could operate in the other direction, with parapsychology being allowed to expand because its work was becoming more acceptable. However, the authors do provide evidence that social position may be used as a resource to strengthen knowledge-claims. For example, they describe instances where parapsychologists' 'proper' scientific training, their affiliation with major research establishments, and so on, were explicitly accepted by the orthodox as factors which required them to take seriously knowledge-claims about paranormal phenomena. But Collins and Pinch furnish no evidence to show that parapsychologists were able to use their improving organisational position directly to ensure the acceptance of any particular knowledge-claim as valid. Thus their central conclusion can best be interpreted as follows: parapsychologists' action in the contingent forum helped to persuade outsiders of the legitimacy of their overall research enterprise; and to create a social context in which particular knowledge-claims, which would previously have been dismissed out of hand, would be more likely to receive careful consideration and, in some cases, acceptance.

So far in this section I have considered only studies dealing with the production of knowledge in new areas of scientific investigation. But a similar combination of social and cognitive processes can be seen to operate in the course of 'revolutionary' situations. To illustrate this, let me look briefly at a study by Frankel (1976) of developments in early nineteenth-century French research into the classical topic of optics. This case is particularly interesting because it shows how variations in the social position of participants may foster the emergence of radically divergent scientific interpretations within a strong and fruitful consensus and also how the social organisation of the research community may affect the success of unorthodox ideas.

The central feature in French optical research during the period 1815-25 was a dramatic swing from the corpuscular to the wave theory of light. Frankel argues that a Kuhnian analysis of this change is not entirely appropriate and by no means complete. From the Kuhnian perspective, we would expect this revolution to have followed a period of crisis within the research network involved; this crisis arising out of an awareness of anomalies defined in terms of the current interpretative framework and widely recognised by participants. However, Frankel shows that in this instance there was no prior crisis and no recognition of major anomalies by adherents of the dominant paradigm until after the revolutionary sequence had already begun. The preceding decade had, in fact, been one of unprecedented success and productive elaboration for those employing the corpuscular theory. At the beginning of the decade of change there was no awareness of important problems which could not be resolved along orthodox lines.

The major change in theoretical perspective which did subsequently occur could neither have begun nor have been brought to fruition without the activities of a group of 'revolutionaries'. This was so because almost all the influential figures in French physics were rigidly opposed to non-corpuscular optics and because these men controlled the dissemination of legitimate information in this field. The scientific revolution in French optics was set in motion by actions carried out in both the contingent and the constitutive forums by a small number of scientists who were isolated, or estranged, from the centre of power and orthodoxy. Frankel identifies Arago and Fresnel as key figures in the revolt. Fresnel was at this time quite outside the research community and eagerly seeking a chance to make a scientific reputation. His sponsor, Arago, was an established scientist, but he was alienated from the ruling clique—until after the success of the wave theory, that is, when he took over the dominant position that Laplace had occupied a decade earlier. Frankel contends that scientists in such marginal social positions are more likely to be receptive to alternative interpretations of results which can always, in principle, be seen in various different ways; and that they are more likely to explore such alternatives because they have more to gain thereby than the established representatives of orthodoxy. It is certainly the case that a large number of major scientific advances have originated with researchers in marginal situations (Chubin, 1976). At the time that Fresnel and Arago began to challenge the corpuscular theory there was one alternative scheme available which had undergone some degree of scientific refinement and which could be used, therefore, to provide immediately fairly sophisticated analysis. This was the wave theory, which had recently been employed by Young in England but which had until then made little impact on orthodoxy in either country.

It is clear that both Arago and Fresnel became committed to the wave theory before any kind of 'proof' of its validity was available, even in their own terms; and that from early on their central aim was conceived as building up evidence in its favour. As their work proceeded, they came to differ increasingly from mainstream scientists in their recognition of anomalies, in the theoretical resources they brought to bear on these anomalies and in their detailed interpretation of data. 'Thus diffraction, for example, was seen as a central problem by Arago and Fresnel, while to Biot it was a relatively minor concern. It is not the nature of the anomaly alone, therefore, but the situations of the scientists involved in the dispute as well, which determines the course of events' (1976, p. 175). Whilst the revolutionaries were attempting to redefine the interpretative framework of optical research, the more orthodox scientists continued to extend the corpuscular theory with undiminished confidence. At least initially, they did this effectively. As Frankel stresses, the corpuscular theory was not at this stage in a state of intellectual crisis. Because most participants continued for some time to accept the corpuscular theory as scientifically adequate, the case for social factors having a decisive influence on the emergence of an opposing perspective seems especially strong.

If scientists' judgements and interpretations are significantly influenced by social as well as cognitive factors, it follows that any deviant or revolutionary group, if it is to succeed, must act at both the social and cognitive levels. Frankel draws attention to three things which such a minority group must do: it must produce 'solutions' to problems regarded as important, it must publicise its views and it must win recognition for its work. The degree to which these objectives can be achieved, he suggests, depends significantly on variations in social context. In order to clarify this point, Frankel makes some interesting comparisons between the position of Young in England and that of Fresnel and Arago in France. The central point is that, as the same *intellectual* resources were available in both countries, the varying success of the wave theory in the two research communities must have been due primarily to social differences. For instance, it was relatively easy for Young to get his ideas published and known in the open, *laissez-faire* community of British science. But he failed completely to convince anyone that his analysis was preferable to the corpuscular theory. In contrast, Fresnel was able to win a hearing from the more closed and centrally dominated community in France only through the 'political leverage' of Arago. However, once this breakthrough had been achieved, Fresnel was able to engage in dialogue with a group of 'competent judges' and to succeed comparatively quickly in establishing that his work could not be ignored.

> The difference in the matter of recognition lay in the state of professionalization of the two communities. Recognition requires agreed-upon standards as to what constitutes an adequate solution to a problem. In the relatively amateurish state of early nineteenth century British physics there was no such agreement, so Young's critics could disregard him for philosophical reasons without confronting his mathematical arguments. In France, the Laplace school and the Ecole Polytechnique had combined to set certain standards which were commonly accepted, even by warring parties in a scientific revolution. If one could predict numerically the results of experiments from theoretical reasoning expressed in the form of mathematical equations, then one was acknowledged to have achieved a solution to the problem at hand. (Not necessarily the only solution, or the best solution, however.) Thus the judges in the diffraction contest could award Fresnel the prize on the basis of his mathematics and his experiments, without committing themselves to his theory. (Frankel, 1976, p. 176)

Consequently Arago, acting on Fresnel's behalf within the councils of the Parisian scientific élite, was able to use established criteria of adequacy as a resource in the informal process of negotiation. He was able to employ agreed standards to define Fresnel's contributions as significant and to convince others that Fresnel's work was consistent with those standards.

Frankel emphasises that, although this recognition of the significance of Fresnel's work was the crucial first step towards the success of the wave theory, it did not settle the issue. Most orthodox scientists continued to deny the validity of Fresnel's interpretations and their opposition was not ended simply by increasingly clear scientific demonstrations in accordance with agreed criteria. The standards shared by both sides were interpreted so as to support quite different scientific analyses. Frankel claims that the rapid rise to dominance of the wave theory in France was not brought about by cognitive factors alone but also by socio-political factors. In particular, the debate over the two theories of light coincided with the victory of an anti-Laplace faction in the French scientific community as a whole. As a result, there occurred a wide-ranging revision of interpretative positions in various fields.

From the early 1820s onwards, supporters of Laplace's view of science, including the adherents of the corpuscular theory of light, were reduced to silence (Fox, 1974). Although many of them refused to recant, they were unable publicly to oppose the new orthodoxies. One consequence of these social changes was a rapid and almost complete acceptance of the wave theory by the Parisian scientific community. From their new positions of power in teaching, research and publishing, adherents of the wave theory were able to convert the next generation of students in its entirety. It seems, therefore, that there was nothing inevitable in this intellectual revolution. It did not occur as a necessary consequence of the logic of scientific ideas, but was set in motion and brought to completion in part by contingent social factors. Where the appropriate social context was lacking, as in England, the corpuscular theory remained dominant for some time to come. It is perhaps relevant to note that this was not the end of the 'corpuscular versus wave' debate, which surfaced again in a new interpretative context in the present century. (See also the discussion of Holton's themata in the next chapter.)

Let me conclude this section by summarising very briefly the main conclusions of the case studies discussed above. I will refer in passing to philosophical sources where similar points are made. In Collins's study of the laser network, we saw that scientific knowledge cannot always be made explicit. It has a tacit component which cannot be assessed by means of formal criteria (see Polanyi, 1958, and Ravetz, 1971). Such tacit knowledge is transmitted most effectively through direct social interaction, in the course of which the adequacy of participants' knowledge has to be informally negotiated rather than formally demonstrated. In the study of gravitational waves we were shown that, even in a research context where the intellectual framework is well established and theoretically elaborated, available knowledge and techniques can be interpreted in quite diverse ways (see Hesse, 1974, and the discussion in the preceding chapter about the variable meaning of concepts, etc.). Beginning from significantly different interpretative positions, participants appeared to be working out, by a process of 'claim and counter-claim', the meaning of observations and thereby, indirectly, establishing the nature of the phenomenon with which they were concerned (see Feyerabend, 1975). As in the case of laser research, this process of negotiation was not confined to the formal channels of communication, but depended to a considerable extent on informal interaction which enabled participants to buttress their formal claims with a wide variety of non-technical evaluations. As we saw in the previous section, the use of such non-technical or social formulations is likely to be highly flexible and context-dependent.

Wynne's study showed how consensus may be achieved and maintained, not by clear disproof of alternative positions, but in part by the highly selective presentation of evidence (see Hanson, 1969, on selective perception), by the 'misrepresentation' of opposing arguments and by attacks on their author's scientific integrity. Wynne suggests that scientists necessarily make strong commitments to research programmes which can never be formally justified in full (see Lakatos, 1970). Like Collins, he stresses the informal, unexplicated, intuitive elements in the creation, justification and maintenance of scientific knowledge. Because the commitments on which the course of scientific development depends cannot be formalised, they are open to influence from a variety of social factors. Wynne concludes that the formal rationality of science is at least partly a form of justification constructed after, and with the help of, social and intellectual commitments. Accordingly, he agrees with Collins that we will obtain a totally misleading view of science if we infer its social attributes from the formal characteristics of the claims presented in articles, reviews and textbooks. Formal knowledge claims have meaning only when they are interpreted by the members of actual social groupings. The way in which these interpretations are realised depends on the outcome of contingent negotiations among those members.

In the case of parapsychology we saw that well-established scientists seem to be able to resist formal constitutive knowledge-claims indefinitely when these claims are based on radically different assumptions (see Kuhn, 1962; Polanyi, 1969). For the critics of parapsychology, the central assumption that paranormal phenomena do not exist was never in question. Rather it pervaded and gave meaning to the whole armoury of formal arguments which they employed (see also Bloor, 1976). It ensured that, for these critics, every item of evidence and every chain of reasoning provided further grounds for rejection of the deviant views. In this case, the informal reasoning and the use of contingent criteria which are normally excluded from open display in the formal arena appeared clearly in the course of knowledge-constituting debate. The authors show how parapsychologists, faced with the apparent impossibility of proving their case through formal discourse yet convinced of its validity, explicitly adopted a supplementary strategy in the contingent forum. They argue that parapsychologists have had some success with this strategy and have been able to use their improved position in the scientific community as a resource for establishing their intellectual claims. This theme is taken further in the final study by Frankel, who argues that scientists' intellectual commitments, their choice from available analytical resources, may be directly affected by their social location; that the success of unorthodox views can be crucially influenced by the internal politics of science; and that the formal rationality of an emergent paradigm may be made unproblematic for a new generation of scientists by becoming entrenched in the centres of scientific authority.

These case studies are not free from interpretative or methodological difficulties. It would be surprising if they were; for studies of this kind have not been attempted until very recently and the practical difficulties of combining an investigation of social relationships with detailed study of intellectual developments are very great (Law, 1976; Woolgar, 1976b). Their conclusions, therefore, should be treated as tentative. They cannot, however, be ignored. At the very least, the evidence produced in these and similar studies demands further analysis and empirical investigation. In the next section I will try to identify the central implication of these path-breaking sociological studies of the production of scientific knowledge and to show that it is supported by the revised analysis of social norms in science. To put this another way, my aim will be to distinguish the central feature of the analytical framework which, strengthened by its consistency with the new philosophy of science, is emerging as an alternative to the customary sociological account of science. But before I do this, I want to make one important point. All the case studies discussed in this section have dealt with fields of empirical science. It should not be assumed, however, that 'non-empirical' disciplines develop quite differently or that their knowledge has a certainty or clarity different in kind from that of empirical science. In other words, we cannot retain the category of 'special case' traditionally used by sociologists of knowledge by confining its scope to mathematics and logic. This has been argued most forcibly by Bloor (1976, chs 6 and 7), who shows that mathematical formulations and logical principles have no meaning until they are interpreted in terms of non-formal, background assumptions; that these assumptions are socially variable; that mathematical reasoning is, therefore, context-dependent; and that mathematical proofs are produced by informal process of social negotiation. Bloor's analysis is easily available and I will not discuss it further. I will, however, assume that the conclusions which follow are likely to apply as much to the formal as to the empirical sciences.

THE INTERPRETATION OF CULTURAL RESOURCES

In the light of the material presented in the two preceding sections, it seems that one crucial fault in the orthodox sociological account of the production of scientific knowledge is the lack of any conception of interpretation or negotiation (Law and French, 1974: Böhme, 1975). This omission, and the gap it leaves in sociological analysis, are evident with respect to the treatment of both social and cognitive/technical resources. From the orthodox perspective, it is assumed that sociologists can identify that set of general normative principles which in practice guides most activities in science and, indeed, which *has* to be institutionalised within the research community in order to guarantee that the great majority of accepted knowledge-claims will be faithful to the real physical world. These general principles are conceived as providing clear prescriptions for virtually all social action involved in the production and certification of scientific knowledge. Their application by participants to particular acts is taken to be quite straightforward. Specific acts are regarded as either conforming to a given rule or not, in a fairly unambiguous manner. I have tried to show, however, that for two main reasons this conception of the normative structure of science is unsatisfactory. First, only part of the range of normative principles actually used by scientists is recognised as playing a significant role in science; and the sub-set selected for attention emphasises unduly those formulations most obviously consistent with the standard philosophical view. Secondly, and more important here, both general principles and more specific operating norms have, as we have seen, always to be *interpreted* in particular cases. In order to depict a given action, whether their own or that of some other, as in accordance with a prescription, participants have to interpret that prescription in relation to supplementary considerations as well as those particular features of the context in which it is being applied that are deemed to be relevant.

Accordingly, I have tried to show that we must revise the established view of the relationship between social norms and the production of scientific knowledge. The meaning of norms is always socially contingent; that is, it depends on interpretation by actors in varying social contexts. Because any specific norm can be made consistent with a wide range of apparently different actions, we cannot regard the production of knowledge as a simple consequence of conformity to any particular set of normative formulations. I have suggested instead that it is more appropriate to treat the norms of science as vocabularies which are employed by members in negotiating meanings for their own and their colleagues' actions. Because scientists have available a considerable variety of formulations, each of which can be applied to individual cases in a flexible manner, any given act can always be interpreted in various ways. The extent to which one interpretation rather than another becomes accepted by participants is the outcome of processes of social interaction or negotiation; that is, as members exchange views and attempt to convince, persuade and influence each other, these views may be modified, abandoned or reinforced. Although social negotiation in science has been little studied as yet, it seems likely that its outcome is influenced by such factors as members' interests, their intellectual and technical commitments, members' control over valued information and research facilities and the strength of their claim to scientific authority.

This argument with respect to social norms in science is supported by a closely parallel argument with respect to cognitive/technical norms. (I use this last phrase to refer to the whole range of research methods, techniques, criteria of adequacy, established bodies of knowledge, and so on, in so far as they are employed as resources for judging knowledge-claims and scientific competence. The similar phrase 'technical norms', which is sometimes used within the orthodox framework, usually has a somewhat narrower connotation.) From the perspective of the customary sociological analysis, the supposed social norms are seen as necessary because they ensure that the technical norms of science will be vigorously applied in the selection of knowledge-claims (see Chapter 1). These technical norms, in turn, are regarded as pre-established, impersonal rules which directly regulate scientists' research actions and intellectual judgements. It is argued that, in this way, the 'entire structure of technical and moral norms implements the final objective' (Merton, 1973, p. 270), that is, the accumulation of objective knowledge. The technical resources of science are seen as being sociologically unproblematic. It is assumed that their interpretation and application will be uniform and independent of variations in social context within the research community. (It is, of course, emphasised that, if the supposed social norms are disrupted, the technical norms will cease to function properly.) However, we have seen in the present chapter that this assumption is unjustified. Not only are *social* norms socially variable, but *cognitive/technical* norms are also open to a considerable range of interpretation in any particular research area. In other words, general evaluative criteria such as 'consonance with established knowledge', 'consistency with the evidence', 'competence', 'replicability', as well as the content of specific bodies of knowledge and technique, all require interpretation in particular instances in much the same way as do social norms (Böhme, 1977). Indeed, it is difficult to envisage how technical resources could be employed differently from social resources in this respect because, as every case study demonstrates, there is no clear separation between the negotiation of social meaning and the assessment of knowledge-claims. *Both social and technical formulations have to be selected and interpreted by participants in particular instances; and both kinds of resources are inextricably combined in the sequence of informal interaction as well as formal demonstration whereby specific knowledge-*

claims come to be ratified. (Thus the distinction between social and technical resources must not be reified. Cognitive/technical formulations are merely one kind of interpretative social resource.)

The sociological analysis of science, then, has previously assumed that the production of scientific knowledge can be explained by showing that general conformity is maintained to sets of formal rules (both social and technical), the strict implementation of which guarantees an undistorted revelation of the real physical world. I have argued, in contrast, that neither of these kinds of rule has a determinate meaning for participants and that implementation therefore requires a continual process of cultural reinterpretation. By means of this process scientists construct their versions of the physical world. The broad similarity between this revised sociological position and the new philosophy of science is clearly evident at this point (see the discussion at the end of Chapter 2). Sociologists and philosophers have converged on a conception of science as an interpretative enterprise, in the course of which the nature of the physical world is socially constructed.

It is also clear, I think, that this view brings science fully within the scope of the sociology of knowledge. I do not mean by this that we can begin to talk of the content of scientific knowledge as 'determined by existential factors'. The kind of analysis implied in such terminology is clearly quite at variance with the view that I have been trying to illustrate and clarify. A rather better general formulation would be that scientific knowledge is established by processes of negotiation, that is, by the interpretation of cultural resources in the course of social interaction. Cognitive/technical resources are employed by scientists in such negotiation; but the eventual outcome depends also on the availability of other kinds of social resources. The conclusions established through scientific negotiation are not, then, definitive accounts of the physical world. They are rather claims which have been deemed to be adequate by a specific group of actors in a particular cultural and social context. There is, then, at least a *prima facie* case in favour of the thesis that 'objects present themselves differently to scientists in different social settings, and that social resources enter into the structure of scientific assertions and conclusions' (see Chapter 1, p. 2).

In the present chapter, I have developed this thesis with respect to processes occurring inside the research community. In the final chapter, I will show that the same approach can elucidate some of the connections between scientific knowledge and the society at large. This will enable me to illustrate how some of the traditional questions of the sociology of knowledge may be applied to the case of science and, perhaps, provisionally answered.

4

Science and the Wider Society

In general, sociologists of knowledge have paid particular attention to the influence exerted by external social factors on the work of specific groups of cultural producers. For instance, Stark (1958) claims that the existence of radically divergent traditions of philosophical thought in Germany and Britain is primarily a consequence of differences in socio-political environment. He suggests also that the transition from classic to romantic art in Europe at the turn of the eighteenth century was fostered by a marked change in the social position of the artist, which was in turn a result of wider social developments. These are typical of the connections between cultural products and society which have been identified by sociologists. Not only the rate and direction of cultural development, but also its content, are portrayed as directly dependent on external influences. In the case of science, however, external factors have been regarded as less powerful. It *has* come to be generally agreed that the speed and direction of scientific development are considerably affected by social, economic and technical factors originating outside the scientific research community (Mendelsohn, 1964; Ben-David 1971). But most philosophers, historians and sociologists have been unwilling to accept that such external factors can influence the *content* of scientific thought, that is, its concepts, empirical findings and modes of interpretation (Young, 1973, Lemaine *et al.*, 1976).

The reasons for this are, I hope, clear by now. Scientific knowledge has been conceived as an objective representation of the physical world. The modern scientific community has been credited with an ethos which reduces social influences upon the production and reception of knowledge-claims to a minimum, thereby guaranteeing the accumulation of objective knowledge. Given these assumptions, direct connections between the wider society and the conclusions of science are not to be expected, except in those few cases, like that of Lysenko, where outside intervention 'distorts' scientists' results (Joravsky, 1970). Consequently, when sociologists have sought to understand the relationship between science and the wider society, they have conceived their central task to be that of demonstrating which kind of society is most amenable to the institutionalisation of the 'scientific ethos' and most likely to support an autonomous research community.

This has led to the thesis that democratic societies furnish the most congenial setting for scientific development, because such societies give academic scientists the freedom they require to record without bias the facts of the natural world and because science and democracy share those values on which the production of valid knowledge depends (Barber, 1952; Merton, 1957, p. 522; Hirsch, 1961; see also Polanyi, 1951). However, the assumptions behind this line of reasoning have been challenged in the two previous chapters. We have seen that it is preferable to think of scientific knowledge as a contingent cultural product, which cannot be separated from the social context in which it is produced. We have also seen that the supposed scientific ethos is merely part of the cultural repertoire of science, and by no means necessarily the most important part of the creation of scientific knowledge. There is no longer, therefore, any reason to expect that science is best created in a social vacuum where institutionalised democratic values allow disinterested researchers to formulate the 'one correct account of the physical world'.

The revisions in the customary view of science which have been presented above enable us to reconsider the possibility of there being direct external influences on the content of what scientists consider to be genuine knowledge. This issue is now empirically open and we can turn to detailed studies of the development of scientific thought to see how far it is influenced, on the one hand, by the actions and cultural products of non-scientists and, on the other hand, by the actions and cultural acquisitions of scientists themselves in non-scientific contexts. We are no longer forced to reject these possibilities out of hand as inconsistent with the cognitive and social character of science. Moreover, although sociologists have not yet explored these possibilities, some social historians of science have begun to do so (MacLeod, 1977). In the next section, therefore, I will examine some recent work in the social history of science. I will not discuss the full range of external influences on science. I will concentrate instead on just a few analyses dealing with the content of science, in order to establish clear links with the argument presented in previous chapters.

SCIENTISTS' USE OF 'EXTERNAL' CULTURAL RESOURCES

We can think of scientists as having access to two main cultural reservoirs: that provided by 'the scientific community' and that provided by the wider society. As the scientific community has grown larger, its own resources have become more extensive. Consequently it seems likely that, as Durkheim suggested, science has become culturally more independent over time. Increasingly it is other scientific sub-cultures that provide interpretative resources, with the products of physics and mathematics being exploited by specialties in chemistry and the life sciences. 'As science has incorporated into itself more and more of the

cultural resources of the societies in which it has thrived, so it has become
more internally self-sufficient, with cross-fertilization between specialties
replacing "fully external" inputs in the process of cultural change'
(Barnes, 1974, p. 119).

Despite this tendency, it is still possible to find external culture being
taken into science. For example, scientists today still make considerable
use of commonsense knowledge which is acquired largely in the course of
non-scientific activities. Close observation of scientists at work shows
that they continually move between a highly esoteric terminology and the
language of everyday life. Thus 'commonsense modes of perception and
operation are an *integral* and *essential feature* of recognized scientific
practice' (Elliot, 1974, p. 25). Even in the most rarified areas of physics,
informal reasoning and debate make use of a wide range of interpretative
notions brought in from ordinary discourse; and not only discourse
about physical objects but also about social relationships. Thus particles
'attract' and 'repel' one another. They are 'captured' and they 'escape'.
They 'experience' 'forces'. They 'reject' and 'accept' 'signals'. They
'live' and 'decay', and so on (Holton, 1973, p. 106). It is, of course, clear
that these terms acquire new meanings as they are used in this unusual
context. Nevertheless, their meaning continues to resemble that with
which scientists are accustomed in the course of their everyday social life.
Physicists adopt these terms because they are familiar and, therefore,
furnish ideas which can be applied by analogy to enable physicists to
reason from the known to the unknown (Deutsch, 1959).

Little is known about such informal processes of reasoning in science
and the kinds of resources which are employed. Only Holton (1973) has
attempted any kind of systematic analysis of historical examples in order
to explore how informal thinking contributes to scientific development.
Like the sociologists whose work was discussed in the previous chapter,
he stresses that informal processes are of fundamental importance in
science and that their significance has not been properly recognised,
largely because participants cover up the transition from private speculation to formal demonstration. When we examine in detail how scientists
actually reach their conclusions, as opposed to the way in which they
present them formally, we are 'overwhelmed by evidences that all too
often there is no regular procedure, no logical system of discovery, no
simple continuous development. The process of discovery has been as
varied as the temperament of the scientists' (1973, pp. 384-5).

Holton's central and well-documented contention is that informal
reasoning in science depends on basic presuppositions which inform and
guide scientists' work, sometimes leading them to 'wrong' conclusions
but also in many cases enabling them to disregard contrary evidence in
pursuit of what is subsequently seen to be the correct interpretation.
Each scientist, Holton argues, makes a commitment to a particular
approach towards his area of study. He commits himself, for instance, to

the notion of atomistic discreteness or continuity, to harmony or conflict, to development or equilibrium, and so on. Holton stresses that these commitments are 'neither directly evolved from, nor resolvable into, objective observation on the one hand, or logical, mathematical, and other formal analytical ratiocination on the other hand' (1973, p. 57). They either precede formal interpretation or they are adopted without formal proof to resolve interpretative problems. Indeed such commitments, of which we have seen several examples above, are not open to direct proof or disproof. Rather they define the point at which it is no longer appropriate to ask further questions. They provide taken-for-granted assumptions which are used to generate interpretations and thus to 'bridge over the gap of ignorance'.

Apart from the evident consistency between this analysis and that developed in previous chapters, the point of particular relevance here is Holton's proposal that this repertoire of basic themes or presuppositions spans the boundaries between the scientific community and the society at large. It is part of a common imaginative inheritance. 'What is interesting is that on certain occasions, during the transformation of conceptions from the personal to the public realm, the scientist, perhaps unknowingly, smuggles the commitment of his individual system, and that of his society, into his supposedly neutral, value-indifferent luggage' (1973, p. 101). Holton shows that in *Greek* science basic presuppositions were taken fairly openly from the social thought of the time. Science then was accepted as an adjunct to moral philosophy. When Holton considers the emergence of modern science he finds that the natural philosophers of Newton's time tried to avoid having to state or discuss the theological and other 'non-scientific' notions which, in practice, contributed significantly to their analyses. This is one of the crucial differences between modern science and its predecessors. The cultural and social roots of knowledge have been hidden away in modern science, on the mistaken assumption that true knowledge should not involve reliance on unverifiable assumptions.

Holton (1973) carries out several case studies to show that the cultural connection between science and society today is not always as different from that of Newton's, or even Plato's, time as we have been led to believe. There is in practice a continual cultural exchange between science and the wider society. Interpretative resources enter science mainly through informal thinking, usually with only a very limited awareness of their external origins on the part of participants. They are refined and modified in the course of informal negotiation; and they are allowed into the public annals of science only after appropriate reformulation. These interpretative resources are not generated by the 'facts of nature', nor by the social life of a segregated research community alone. They must be understood at least in part as products of the social processes of society at large.

If this analysis is correct, it should be possible to produce two broad classes of supporting evidence. In the first place, it should be possible to discern parallels between the style of thought in certain areas of science and that occurring in other areas of cultural production, such as painting or philosophy—areas which are clearly influenced by the surrounding social context. Holton has begun to do this, but his evidence is as yet quite tentative (see also Kroeber, 1944). The second kind of evidence is that provided by detailed historical study of specific scientific developments. I intend now to look at the emergence of the Darwinian theory of evolution where, because a major scientific upheaval was involved, the documentation is extensive and the movement of ideas relatively easy to discern.

Both sociologists and historians have generally treated the content of the Darwinian theory of evolution and its acceptance as being independent of the social setting in which it was brought forth. We saw in Chapter 1 how Stark argues that Darwin's theory is simply a summary account of observable facts and is not, therefore, open to sociological analysis. Historians similarly have distinguished Darwin from other evolutionary writers, such as Lamarck, Chambers and Spencer. It is accepted that the speculations of these authors, which seem now to have been mostly inaccurate, were influenced by ideological and other non-scientific factors. But Darwin is seen as the first to recognise and describe the actual mechanism of evolutionary change. Consequently, his analysis is presented simply as a detached response to objective evidence and 'is treated in relative isolation from the social and intellectual context in which he worked' (Young, 1971a, p. 443). Let me try to show just how misleading this view is.

Five important elements in Darwin's theory can be clearly identified. The first of these was the belief that the facts of natural history, comparative anatomy, paleontology, and so on could be explained better by a conception of evolutionary development than by the traditional notion that species were stable and had been created more or less in their present form. The second element was the attempt to show that species did in fact change over time. Because it was impossible to obtain systematic evidence of such changes as they occurred (presumably) in the wild, Darwin turned to the close observation of domestic animals and plants. The third crucial element was the assumption that different biological structures were functionally adapted to different kinds of environment. Fourthly, Darwin saw a direct parallel between the process of adaptation in domestic organisms and that taking place in natural settings. The *artificial* selection of certain biological forms among domestic animals and plants, in accordance with the preferences of breeders, provided a model for understanding the *natural* selection which occurred, in accordance with the requirements of survival, in the wild. Lastly, Darwin accepted that the world of living things operated in a uniform fashion. There were, therefore, universal regularities in the realm of biology as well as in the domains of astronomy and physics. Thus Darwin regarded his account of the mechanism of 'natural selection' as expressing a law of nature. Let me examine the origins of each of these elements.

There can be little doubt that theories of evolution sprang up in the late eighteenth and early nineteenth centuries in those countries where capitalism was most advanced, that is, in Western Europe and particularly in Britain. Sandow (1938) has shown clearly that the notion of gradual and continuous biological evolution occurred in those societies at that time as a response to the massive accumulation of new information about plants, animals and fossils. This information had accrued as a by-product of such developments as the worldwide exploration in search of markets and the growth of the mining industry. Incidentally, then, the economic and political expansion of capitalist Europe generated evidence which led to the formulation of new biological theories. The new data obtained haphazardly by men in pursuit of practical affairs often appeared to be at variance with established biological views and, for some scholars at least, implied the need for a radically new interpretative approach.

By the fourth decade of the nineteenth century a new kind of specialism had emerged, which included Darwin, Huxley, Hooker and Wallace. The members of this specialism had obtained first-hand knowledge of biological variation by taking advantage of the opportunities for doing field work offered to wealthy amateurs by the official voyages undertaken to improve trade routes and to consolidate colonial empires. Virtually all of these men came to believe in the reality of biological evolution. Thus Darwin's acceptance of the general idea of evolution was made possible by his social position; that is, by his belonging to a society which had access to a wider range of biological evidence than ever before, by his having sufficient income to devote himself entirely to science, and by his membership of a distinct sub-culture which had already produced several evolutionary theories. But what of the content of Darwin's work? If we are to understand this in greater detail we must do more than link Darwinian theory with broad features of nineteenth-century capitalism. We must seek the social origins of the more specific elements of Darwin's analysis.

Darwin was unique among biologists of his day in devising a long-term programme of recording the details of variation in plants and animals under domestication. This was his solution to the problem of showing clearly that changes in biological structure do occur and that they can be produced by selective inheritance. In pursuit of this objective, Darwin 'collected facts on a wholesale scale ... by printed enquiries, by conversations with skillful breeders and gardeners and by extensive reading' (Darwin in Sandow, 1938, p. 321; see also Vorzimmer, 1969). Most of his observations on domestic variation were, in fact, taken directly from the work of breeders (Young, 1971a); and in order to obtain this information Darwin spent a great deal of time mixing with businessmen, commercial breeders and fanciers. There can be no doubt that Darwin's treatment of domestic variation was 'rooted in the practical activities of plant and animal breeders'; activities whose success was measured, not by the validity of members' knowledge, but by the amount of their financial profit (Sandow, 1938, p. 332). The assumptions and actions of these practical men were absorbed by Darwin. They entered into his scientific assertions and they provided him with criteria for warranting his own knowledge-claims. For instance, in clinching his argument that selected inheritance is the source of variation in domestic animals, Darwin refers to the fact that 'breeders of animals would smile' at any contrary opinion. He goes on to cite several cases where large profits were obtained by selective breeding and he finishes with the statement that 'Hard cash paid down over and over again is an excellent test of inherited superiority' (Darwin in Sandow, 1938, p. 322).

It appears, then, not merely that Darwin's work was made possible by the high level of attainment reached in commercial breeding in nine-teenth-century England as well as by the other aspects of capitalist development already mentioned, but also that his observations, conclu-sions and criteria of adequacy in relation to *domestic* variation were in some degree taken over from commercial breeders. The perspective of the breeders guided Darwin's detailed reasoning about domestic varia-tion and, thereby, his inferences about the importance of selective inheritance as the source of evolutionary adaptation. But the significance of Darwin's close relationship with the breeders does not end there, for their procedures also provided the central metaphor or interpretative theme which informs the rest of his evolutionary theory. I will return to this metaphor shortly. However, in order to understand the sources of Darwin's analysis more fully, we must digress briefly to consider several other notions which he took over from theological and philosophical debate about the future of society and man's place in nature (Young, 1969).

Darwin's use of such notions is evident in his treatment of adaptation. The variation found in domestic plants and animals is adaptive in the sense that particular forms are selected by breeders in accordance with their own purposes and the demands of the market for which they cater. It does not follow necessarily from this that variation in natural settings is also adaptive. Darwin assumed, nevertheless, that domestic and natural selection are equivalent in this respect. He adopted the 'tacit assumption that every detail of structure, excepting rudiments, was of some special, though unrecognized service' in natural as well as in artificial environments (Darwin in Young, 1971a, p. 468). There seem to have been two sources for this presupposition. One was William Paley's writings on natural theology, which were intended to show that all features of the natural world were designed by a beneficent God. Darwin admitted that he was not able to annul the influence of this belief, which was then very widely held. In Darwin's work the idea that God actively constructs each species to a preconceived design is abandoned; but the idea that every item of biological structure plays a useful function is retained.

The second source was Lyell's philosophical principle of the uniformity of nature. Darwin used this notion to argue that the processes underlying variation in natural and domestic settings must be identical; although, of course, the means whereby specific variations are selected certainly differs. It is important to realise that there is nothing inherent in the principle of uniformity which requires us to accept that domestic and natural variation are in the same domain of phenomena. The principle in no way specifies the range of its own application (see Chapter 2). This is illustrated by the fact that it took Darwin and Lyell many years to agree that human beings came within the range of phenomena covered by evolutionary theory. Lyell objected to the idea that uniformities applicable to animals also applied to human beings. It is clear then that, although Darwin sometimes claimed to be following a 'true Baconian method' and simply to be collecting facts, he actually employed presuppositions taken from philosophical and theological debate, and used them selectively to extend and define the scope of his scientific analysis. Indeed this point can be taken much further. For virtually every interpretative notion used by the biological evolutionists had been employed earlier in the course of debate about society and human progress (see Young, 1969 and 1971a, for documentation of this point).

Once Darwin had concluded that domestic and natural variation are equivalent, he completed his theory by extending his interpretation of domestic breeding to species in natural settings. In doing this, he retained to a surprising degree the terminology of intentions and purposes appropriate to the actions of commercial breeders (Young, 1971a). Even in the formal presentation of his theory, he asked his reader to conceive of natural selection as being carried out figuratively by a 'being infinitely more sagacious than man' who operated on biological populations in the wild so as to select 'exclusively for the good of each organic being' (Darwin in Young, 1971a, p. 461). This metaphor continues throughout Darwin's exposition.

> It may be said that natural selection is daily and hourly scrutinizing throughout the world, every variation, even the slightest; rejecting that which is bad, preserving and adding up all that is good; silently and insensibly working, whenever and wherever opportunity offers, at the improvement of each organic being in relation to its organic and inorganic conditions of life. (Darwin, 1859, p. 84)

The metaphor of natural selection, which implied that there was an agent who did the selecting, caused Darwin many problems. For the rules of scientific interpretation developed in the seventeenth and eighteenth centuries had sought to banish anthropomorphism from the scientific repertoire. Indeed, it was partly on these grounds that previous evolutionary schemes, such as that of Lamarck, had been found wanting. Accordingly Wallace, Lyell and other fairly sympathetic commentators criticised Darwin for using this kind of terminology and for thinking 'unscientifically'. But Darwin, although he effectively rewrote the whole original text of *The Origin of Species* in subsequent editions, did little to reduce this element in his analysis. Darwin's refusal to change his form of presentation is particularly surprising because it created confusion over what he recognised to be 'the only novelty' in his analysis, that is, his account of the mechanisms of natural selection. Whereas Darwin professed that 'natural selection' was a metaphor for objective, impersonal laws of nature, many of his readers took the phrase literally. One reason given by Darwin for none the less retaining the term 'natural selection' was 'that it is constantly used in all works on breeding' (Darwin in Young, 1971a, p. 464). This justification illustrates Darwin's strong commitment to the perspective of the breeders. But there must have been other considerations involved, to account for his general use of a voluntaristic vocabulary and his unwillingness to drop his metaphor once it had been shown to confuse rather than clarify his argument. One such consideration was that Darwin's metaphor enabled him to avoid having to *demonstrate* that domestic and natural variation were equivalent. For this demonstration was impossible. The evidence available on natural variation was insufficient to establish the connection. Thus Darwin used the metaphor of natural selection and a terminology derived from the work of breeders to bridge over a major gap in his argument. He began with artificial selection and familiar examples in order to convince his readers that biological forms could be selected in accordance with external requirements. He *assumed* that in natural settings biological forms were functionally adapted. And he used the language of domestic selection in his analysis of natural contexts in order to bring his readers to make the 'leap of faith' required to see natural and domestic adaptation as equivalent processes (Young, 1971a).

I have suggested, then, that the form of argument used by Darwin in his treatment of natural selection was a consequence of his commitment to the anthropomorphic perspective of commercial breeders and of his inability to provide detailed empirical evidence to support a thesis of which he was personally convinced but which he was unable to demonstrate formally. It follows from this latter point that Darwin's account of the mechanism of natural selection, the linchpin of his theory, cannot be treated simply as a summary statement of regularities observed in nature. Let us look briefly at how Darwin, and Wallace who independently reached the same conclusion, arrived at the 'laws of natural selection'. In the first place, it is clear that their formulation was not gradually achieved by the systematic accumulation of evidence; although it *was* preceded by much data-collection. Both Darwin and Wallace had long had enough evidence to be convinced that natural evolution occurred and were looking for an explanation, when they each experienced a dramatic flash of insight to the effect that Malthus's 'laws' governing human populations could be applied with even more force to animal and plant life. As a result, they suddenly saw the corpus of available observations from a new perspective.

There is clear evidence here of informal, indeed subconscious, thinking; of the kind of *gestalt* switch emphasised by Hanson, Kuhn and Holton. In this case, a solution to a major interpretative problem in biology was suddenly seen to be solvable by the displacement of concepts from another area of cultural activity (Schon, 1963). In order to understand this transfer of ideas, it is important to realise that Malthus was the most widely discussed author in Britain in the early years of the nineteenth century. Malthus's analysis served as a basic cultural resource for numerous areas of intellectual life at that time. Moreover, not only was his argument employed by all the major evolutionary writers, but it was used in a variety of different ways. Malthus himself argued that the growth of *human* populations would prevent social progress from continuing beyond a certain level. Paley used the same argument to show that *biological* species were stable. Lyell drew upon Malthus to demonstrate how species were *eliminated*. Whilst Darwin and Wallace used the same basic idea to account for the *origin* of species. Malthus had undoubtedly hit upon a powerful, multi-purpose interpretative formulation. There was something about the Malthusian doctrine which fascinated the educated nineteenth-century British mind and which expressed what Holton would call 'the style of thought of the age'.

Malthus's thesis grew out of the late eighteenth-century debate about the nature of man and the future of society. Utopian writers like Godwin and Condorcet had argued that man is infinitely perfectible, that human reason is supreme and that complete social harmony is attainable. This conception of social progress was a dominant eighteenth-century theme. Malthus argued, in contrast, that human and social perfectability were limited by the operation of a basic law of society; namely that whereas population tends to increase geometrically, the means of subsistence increases only arithmetically. It follows that population growth will always outrun subsistence and will be kept in check only through the elimination of the 'poor and inept' by the ruthless agencies of hunger and poverty, vice and crime, pestilence and famine, revolution and war. This central part of Malthus's analysis was taken over by Darwin and by Wallace, and applied to the selection of biological organisms in natural settings. The best-adapted biological forms were seen as surviving the struggle for life in the wild, in exactly the same way that the fittest individuals were thought to survive the rigours of industrialisation in *laissez-faire* Britain or equivalent pressures in other societies.

Malthus's original argument was so influential because it made sense of some of the disturbing social changes accompanying industrialisation and because it explained these events as the inevitable consequences of natural law (Sandow, 1938). Many members of nineteenth-century society were engaged in a bitter struggle for existence. Malthus sought to show that this had always been so and always would be so. Thus what Darwin and Wallace did was to transpose into the domain of biological theory an interpretation which had received its impetus and its justification from consideration of social phenomena. However, they carried out this transposition without having the detailed evidence necessary to document the biological laws which had to be assumed to operate if the Malthusian interpretation were to be accepted. Darwin was already convinced of evolutionary change and of the equivalence of domestic and natural adaptation. What he needed was a formulation which could explain adaptation in natural settings and provide a parallel to the actions of breeders on domestic populations. As he had no detailed evidence on which to build up a picture of how 'natural selection' worked (it could be presumed that this lack of evidence was due to the fact that the process was so slow and because the fossil record was fragmentary), Darwin completed his analysis by adopting the ready-made thesis available in Malthus. But although the Malthusian doctrine could be redesigned to meet Darwin's requirements, it provided no more than a general interpretative formula. It is hardly surprising, given its origins in social debate, that Malthus's argument did not bring Darwin any closer to specifying, for example, how inheritance is transmitted or how particular structures cope with environmental pressures.

> In proposing the theory of evolution by means of the mechanism of natural selection he was not really supplying a mechanism at all. Rather, he was providing an abstract *account* at a general level of how favourable variations might be preserved. He *had* to keep his account at a certain level of abstraction since, as he confessed, he could neither specify the laws of variation nor the precise means by which variations were preserved. The acceptability of his account depended on its plausibility and its ability to explain in very general terms the *sort* of process which was involved. (Young, 1971a, p. 469)

Darwin, then, became convinced informally of the parallel between the selective procedures of breeders and the effects of Malthusian pressures on species in the wild; but he could not show in detail how artificial or natural selection worked. In the formulation of his theory, therefore, he covered over the gaps in his knowledge of Malthusian natural laws, as he covered over the assumption that domestic and natural variation are equivalent, by use of the anthropomorphic metaphor and the device of the figurative 'wiser being'. Yet, despite the lack of evidence, the metaphorical argument and the continual weakening of his major claims in response to criticism, Darwin's theory was immensely influential and highly successful in gaining widespread acceptance of the general idea of organic evolution. Two factors seem to have contributed significantly to this success. On the one hand, as Young (1971a) has shown, Darwin's voluntaristic terminology made it easier to forge a theological position which could encompass his conclusions. God came to be seen as the 'wiser being' who had designed the laws of nature so as to ensure the beneficent adaptation of biological forms. But equally important was Darwin's reliance on the Malthusian argument which Victorians found so persuasive. As we have seen, the only *novel* part of Darwin's theory was his application of Malthus's argument to account for the emergence of new species (although Darwin's research technique of acquiring data from breeders was also exceptional). It appears, therefore, that it was the combination of the evolutionary hypothesis with the powerful Malthusian thesis purporting to show how evolution actually worked, which made Darwin's argument so convincing to many and a source of such dismay to others.

The heavy reliance on Malthus by Darwin, and to a lesser extent by other evolutionists, is particularly interesting because it seems to have been based on certain assumptions about the nature of their society. This can be seen most clearly in the case of Wallace. In the late 1870s, Wallace read and was convinced by Henry George's socialist interpretation of society. Wallace came to believe that voluntary co-operation and reform were important social forces and could replace struggle and competition as the central agencies of social change. As Wallace changed his view of the basis of social life, so he rejected Malthus's analysis. Malthus's theory, he came to see, had no bearing 'whatever on the vast social and political questions which have been supported by reference to it' (Wallace in Young, 1969, p. 133). This seems to show that Malthus was convincing only so long as it was assumed that the marked inequality, the ruthless competition, the absence of welfare legislation, and so on, which were characteristic of nineteenth-century Britain, were *necessary features of society*. Only on this assumption was it possible to treat Malthus's formulation as a natural law of society which was sufficiently well established to be transposed to the biological realm. There are some grounds, therefore, for suggesting tentatively that the use of Malthusian doctrine by Darwin and his colleagues was made possible by their sharing a series of background assumptions about the nature of social life which were derived from dominant features of their own society; and that similar presuppositions contributed to the influence of Malthus's writings and to the success of evolutionary theory.

The case of Darwinian theory seems to illustrate most of the main points made by Holton. Fundamental scientific conclusions were reached by informal, metaphorical procedures of reasoning which, despite strenuous attempts at formalisation, left distinct traces on the form and content of the ensuing knowledge-claims. Scientists drew intuitively on external interpretative resources which were used to define the nature of their intellectual problems and to fill in the gaps in their analysis. The major interpretations proposed were not based directly on observation of biological phenomena. Rather they were taken over from practical activities and from the wider realm of philosophical, theological and social debate to provide the framework within which observations were given their scientific meaning. During the period that Darwinian theory emerged there seems to have been a body of related ideas, associated with the theme of struggle and adaptation, which was the main interpretative resource for a wide variety of intellectual endeavours. Whenever these were adopted by members of the emergent biological research specialism, efforts were made to separate them from their social origins and to devise formulations in which, as far as possible, they could be presented simply as descriptions of observed regularities (Young, 1971a). Consequently background assumptions originating in specific social relationships (as between Darwin and the breeders) or in broader features of society (like those implied in the Malthusian doctrine), have been gradually hidden from sight. As a result, most sociologists and many historians have accepted at face value Darwin's own Baconian account of the epistemological status of his conclusions. They have therefore failed to realise just how fragile the observational basis of Darwin's theory was and how much acceptance of his claims depended on commitment to shared presuppositions. Given that Darwin's theory influenced so many aspects of modern biology, it may well be that some version of these presuppositions has become built into the technical culture of biological science (see Rose and Rose, 1976, ch. 6).

In the pages immediately above I have looked at one case in sufficient detail, I hope, to be convincing, instead of skimming more superficially through a number of instances. The critical reader, however, may think that too much reliance has been placed in this section on the one example. For the case of Darwinian theory might be misleading in two respects. First, as I pointed out above, the scientific community has probably become increasingly able to satisfy its own cultural requirements during the last hundred years or so; and biological research, in particular, has undoubtedly become more specialised, more technical and less closely linked to broad social debate than it was in Darwin's time. Thus the Darwinian illustration may seem out of date. Secondly, it may be that nineteenth-century biology was more open to outside influence (or less separated from what we now see as distinct areas of discourse) than has ever been true of physics and chemistry, which are further removed intellectually from the social realm. There is a modicum of truth in both these propositions. The academic research community *has* been able to achieve a very considerable autonomy and social segregation; and the connections between scientific thought and social life have tended to be more obvious in the case of biology than in the case of chemistry and physics. Nevertheless, in recent years a number of historical studies have been completed which show that external influences on the content of science have been confined neither to the distant past nor to the life sciences (see also Young, 1971b, and Rose and Rose, 1976, on modern biology). Let me simply list some of the fields for which a reasonable case has been made and refer the reader to the sources given: thermodynamics and the conservation of energy in nineteenth-century physics (Brush, 1967; Elkana, 1974); the theory of relativity (Feuer, 1971; Holton, 1973); quantum theory (Forman, 1971; Holton, 1973); and present-day organic chemistry (Slack, 1972).

We can be fairly certain, then, that the analysis of Darwinian theory does exemplify processes which occur in the 'harder' sciences and in the present century. It seems, therefore, to be possible to supplement the account sketched in Chapter 3 of the processes of knowledge production inside the research community with a more macro-sociological analysis (this kind of general approach is advocated in Stark, 1958); although, of course, much more comparative analysis is needed than can be attempted here. One way of linking internal and external processes may be through the notion of 'interpretative failure'. In other words, it seems likely that scientists turn to other cultural areas when basic interpretative problems prove particularly difficult to resolve with their existing resources. In such situations, scientists are likely to look beyond their own community to other relatively systematic and co-ordinated bodies of analysis or to coherent practical traditions. Although interpretative failure seems to have been involved in the case of quantum theory as well as Darwinian theory, it does not always seem to be present. The crucial factor in the case of organic chemistry appears to have been the longstanding relationship between the research community and the chemical industry. This is a useful reminder that external demands expressing group interests may influence how scientists conceptualise their field, irrespective of external cultural resources (see also Nelkin, 1977, for a discussion of ecology, and Johnston, 1978, for a more general view). Moreover, bodies of culture are carried by social groups and the movement of cultural resources is often mediated through the relationships between collectivities. (This is probably least true of that commonsense knowledge which plays an as yet mysterious role in scientific thinking.) However, I have by now established the general point that the content of science is affected by social and cultural factors originating outside science. In the next section, I shall consider the movement of cultural resources in the opposite direction, that is, from the scientific community into the wider society.

We will find that structured social relationships and group interests are crucially important in understanding the processes involved.

THE USE OF SCIENTIFIC CULTURE IN EXTERNAL POLITICAL SETTINGS

One major concern in the sociology of knowledge has been to show how society at large influences the production of specialised knowledge. In the last section I tried to show that science need not be exempted from this kind of analysis. A second important objective for the sociology of knowledge has been to explore how knowledge is used in the course of political activity. I intend now to consider briefly this second theme. Sociologists' accounts of science and the political process have been formulated in accordance with the customary assumptions about the character of scientific knowledge and the nature of the scientific ethos (see Blume, 1974). As a result, science and scientists have been seen as occupying a rather special position in political life.

In the first place it has been assumed, with little recognition of the need for close empirical documentation, that scientists are the bearers of a type of knowledge which is unaffected by the social context in which it is used. 'The theories, models, procedures and formulae of science . . . are generally believed to allow one trained in their use simply to calculate an unambiguously correct answer' (Mazur, 1973, p. 251). Secondly, it has been accepted that the supposed norms of science, such as disinterested-ness and universalism, require scientists to act in a politically neutral manner and that scientists continue to abide by these norms outside the research community (Barber, 1952; Brooks, 1964; see also Ezrahi, 1971).

Clearly, if these two assumptions are correct, the increasing participa-tion of scientists in the political sphere is likely to change dramatically the character of the political process (Lakoff, 1977); that is, the realm of 'politics and ideology' will be reduced, and perhaps eventually elimina-ted, as the range of application of certified scientific knowledge is extended. Thus 'if one thinks of a domain of "pure politics" where decisions are determined by calculations of influence, power, or electoral advantage, and a domain of "pure knowledge" where decisions are determined by calculations of how to implement agreed-upon values with rationality and efficiency, it appears . . . that the political domain is shrinking and the knowledge domain is growing . . . ' (Lane, 1961, pp. 61-2; see also Bell, 1960). Given the traditional sociological conception of science, this line of argument, sometimes called the 'end of ideology' thesis, is quite reasonable. The assumptions on which it is based are, however, clearly inconsistent with the approach which I have been developing. Let me try to show, therefore, where the argument is inadequate and how we can depict more accurately the way in which scientists employ their cultural resources in the political context.

In the traditional analysis of science and politics which has led to the 'end of ideology' thesis, scientific culture enters in the form of the standard set of social norms and in the form of context-free knowledge. Let me discuss the social norms first. These norms have typically been conceived, as we have seen, as a set of rules unambiguously specifying certain kinds of social action. In the realm of political analysis, they have been interpreted as requiring scientists to adopt a disinterested, politically neutral, concern for objective data. Active engagement in politics has been considered as alien to scientists and as 'essentially destructive of scientific endeavour' (Haberer, 1969, p. 1). However, we have seen that the normative resources of the scientific community are by no means limited to those which have been customarily accepted as defining the scientific ethos. Furthermore, we have seen that the norms of science are best conceived, not as clear prescriptions specifying certain kinds of action, but as standardised verbal formulations from which scientists select in order to establish interpretations of their actions appropriate to particular social contexts. There are now a number of historical studies available of the connections between science and government, particularly in the USA, which are consistent with this revised view of scientific culture. They show that when scientists have entered the political context they have drawn selectively on their cultural repertoire in a way which has furthered their collective interests. They have used their resources to construct a political image or ideology particularly suited to the American 'democratic' setting.

In the first half of the nineteenth century, American scientists did not present a uniform image of science in their dealings with laymen. This seems to have been because they were involved with a variety of lay audiences and as a result produced varying accounts of what science was, what science ought to be and what part science played in society. As the century progressed, however, and as the scientific community increasingly emerged as a separate social entity, a more coherent view began to crystallise.

> Previously, science had been 'sold' to the public in terms of its contribution to important American values—utilitarian, equalitarian, religious—or even as a means of social control, depending upon the speaker's best estimate of his audience. But in the 1870's for the first time, great numbers of scientific spokesmen began to vocally resent this dependence upon values extraneous to science. The decade, in a word, witnessed the development, as a generally shared ideology, of the notion of science for science's sake. (Daniels, 1967, p. 1699)

This ideology was further elaborated and formalised, over the years, particularly in the course of political negotiation over the provision of support for science.

From the late nineteenth century until the present day, one of the crucial factors influencing scientists' relations with government has been the cumulative increase in the size of the scientific community and in the cost of research. Increasingly, scientists have come to recognise that only central government can provide funds on a scale sufficient to maintain the pursuit of scientific knowledge (Price, 1963; Sklair, 1973). At the same time, scientists have striven vigorously, and with considerable success, to maintain what Gouldner calls the 'functional autonomy' achieved by academic or 'pure' science during the last century. In the course of their negotiations with government, scientists have argued for, and have been granted, extensive and increasing support in terms of research funds, educational facilities and personnel, coupled with minimal regulation from outside—particularly with respect to control over the élite activity of pure research. The arguments employed by scientists in these negotiations have come to take a fairly standardised form (Greenberg, 1969; Tobey, 1971; Haberer, 1969; Weingart, 1970). It has been argued, not only that scientific knowledge is intrinsically valuable, but also that, because it is the only truly valid type of knowledge, it necessarily leads to practical benefit. Science is depicted as being unique in its cumulative acquisition of unquestionable facts; which are obtainable only so long as scientists are allowed to approach the study of nature with values which curb human tendencies towards bias, prejudice and irrationality. These values are described by scientists in terms, such as independence, emotional discipline, impartiality, objectivity, a critical attitude, and so on, which are virtually identical to those used in the customary sociological analysis. Moreover, the parallel does not end there. For scientists themselves had argued explicitly, several decades before sociologists of science did so, that science and democracy were especially compatible owing to their pursuit of common values and their common recognition of the need for scientific autonomy (Mulkay, 1976a). For example, during the 1920s and 1930s in the USA, an influential scientific pressure group actively sought to gain acceptance within government and among the general public of the notion that 'American democracy is the political version of the scientific method' (Tobey, 1971, p. 13).

When sociologists first began the empirical study of the scientific community, they seem to have taken such public pronouncements by scientists more or less at face value and to have modelled their own analysis upon them. Sociologists seem to have been noticeably lacking in scepticism towards the public statements of scientists, compared with those of other groupings—probably because sociologists accepted without question the covert epistemology from which scientists' accounts drew much of their strength. In the light of the analysis in previous chapters, however, we can see that scientists' typical account of the nature of science and scientific values is quite inadequate. We cannot accept, therefore, that these standardised formulations were used by scientists in the political context simply because they were accurate or were the only ones available. Given that scientists could have derived quite different, yet still entirely plausible, accounts from their cultural repertoire, it seems reasonable to conclude that this particular version was chosen because it helped scientists to achieve their central objectives in the political context of North America. For if scientists select descriptions and justifications from the available vocabulary in accordance with their interests and the nature of the social context *within* their own community, that is, when dealing with persons who have first-hand experience of the social world of science, there is every reason to expect that they will do likewise in the course of their interaction with laymen, who will find it even more difficult to challenge their accounts. It is certainly clear that scientists have *used* their standard portrayal of science to justify political claims (Greenberg, 1969). Thus the epistemological element has been used to justify increased support for science: 'Science must be kept healthy because it is the only source of valid knowledge.' And the description of the supposed values of science has been used to justify its continued autonomy: 'Science has a clear code of ethics which will only be disrupted by outside intervention. Such intervention can only lead to the distortion of scientific results and eventually to economic and military decline.' This selective characterisation of science, by scientists, in the political context amounts to the creation of a professional ideology.

So far in this section I have offered no more than the briefest sketch of the implications for the political analysis of science of my previous discussion of its normative culture (for further analysis, see Mulkay, 1976a). If this interpretation is broadly correct, it means that the special compatability of science and democracy is little more than a conception created by scientists themselves for their own practical purposes. It also means that the 'end of ideology' thesis does not apply when scientists are involved in political negotiation about science itself. In this latter context, scientists seem to resemble other groups in pursuing their sectional interests and in developing an ideology in order to improve their chances of success. But what of situations where scientists' own interests as a community are not directly involved and where scientists take part merely as providers of certified knowledge? It appears to follow from earlier arguments about the context-dependence of knowledge-claims that scientists' use of their technical repertoire will not differ in principle from their use of social formulations; that is, the way in which scientists interpret and draw on their expertise outside the research community will vary with the social setting in which they are operating and with their position in that setting.

In recent years several detailed studies have been carried out of the use made of scientific knowledge in the course of political debate. The main conclusion of these studies is that scientific knowledge does not reduce the scope of political action, but rather it becomes a resource which can be interpreted in accordance with political objectives. This is clear, for example, in Nelkin's study of the Cayuga Lake controversy (1971; 1975). The central scientific issue in this case was the environmental effect of building a nuclear power plant at Cayuga Lake. The State Electric Company, after it had been challenged on this issue by local scientists, spent over 1½ million dollars on research. The recipients of this money produced results which were interpreted as providing clear refutation of previous criticisms and as entirely vindicating the company's original plan. Various local groups, however, still refused to accept the company's proposal, raised new problems and continued to produce alternative scientific analyses. Participants were no more able to agree about the 'scientific realities' after five years of debate than they had been at the outset and the dispute ended with the political defeat of the Electric Company, despite its massive expenditure on research.

The use made in this case of scientific knowledge-claims seems not to be unusual; nor is the selective use of such claims confined to issues which fall within the domain of 'immature' scientific disciplines (King and Melanson, 1972; Nelkin, 1975). Opposing parties in political disputes involving technical issues can usually obtain the services of reputable scientists who will provide data to buttress their policy and to undermine that of their opponents (Benveniste, 1972). By now it is clear why this is possible. We have seen that the formulation of scientific facts depends on prior commitments of various kinds, that these commitments are often made in accordance with participants' position in a specific social setting, and that they influence the informal acts of interpretation which are essential to give meaning to observations. We would expect, therefore, that scientists occupying differing positions in a political context would often bring different presuppositions to bear and that their informal reasoning would be subtly influenced by the assumptions of the group to which they were affiliated. This view receives support from the few studies at present available. These studies indicate that scientists' entry into the political arena affects their interpretation of their technical culture in three ways: it influences their definition of technical problems; it influences the choice of assumptions introduced in the course of informal reasoning; and it subjects scientists to the requirement that their conclusions be politically useful.

The selective definition of problems is particularly obvious in the Cayuga Lake controversy. In this case scientists defined the technical issues in various different ways which, although they may have been reconcilable in principle, were not in fact reconciled during the debate but were instead used to generate different technical programmes, to 'justify' different policies and to support different interests. For example, in one scientific report, sponsored by the Electric Company, the scientific problem was defined as involving only the lake drainage area and the conclusion was reached that the environmental effect of the power plant would be negligible. Some scientists in the local community, however, maintained that the total parameter of the lake had to be considered and that when this was done the conclusions of the company-sponsored research about thermal pollution had to be rejected. Other scientists thought that *this* was too restricted a framework and urged that the input of the power station be considered as part of the total lake system. This perspective, once again, gave rise to a different scientific analysis and to different practical implications. Mazur's (1973) examination of the debates over the effects of radiation and fluoridation upon public health illustrates the same tendency for scientists to define issues differently and consequently to reach divergent conclusions. Mazur shows that in both these debates some scientists concerned themselves primarily with *acute* poisoning and accordingly judged the risks from radiation or fluoridation to be low; whilst others thought in terms of *chronic* poisoning, the risk of which appeared to be significantly higher. As in the Cayuga Lake example, the lengthy and critical nature of these disputes as well as the failure to reconcile differences shows that scientists became remarkably committed to these narrow definitions of the problem.

Scientists' choice of a particular definition of a technical problem cannot itself be decided by observation and systematic inference alone. Rather, it precedes and is presupposed in observation and analysis. Moreover, the choice of such a definition in these political debates often seems to have depended on a prior social commitment (see Nelkin, 1971, and Mazur, 1973). Thus in the studies already mentioned it seems that those scientists who spoke on behalf of the Water Authorities, the AEC or the Electric Company, chose perspectives which defined the issues in a manner favourable to their patrons' policies. Similarly, those who represented opposing interests worked from quite different definitions which enabled them to reinterpret and to challenge their adversaries' conclusions, and to defend their own collectivity from what they saw as a threat.

Mazur notes that it is possible to treat these disputes as arising from failures of communication which could perhaps be remedied by showing both parties that they were defining the problem differently and that they were, therefore, arguing at cross-purposes. He stresses, however, that this seldom, if ever, happens in practice. It appears to be very difficult for scientists engaged in public debate to adopt a Mannheimian strategy and to redefine their problems at a higher level where apparently divergent views can be reconciled. It is much more usual for differences in underlying definitions and in basic premises to become obscured or to be ignored as both sides present their findings as 'the definitive facts' about radiation or whatever. Mazur and Nelkin have shown that the various parties in any particular dispute tend to use an identical pattern of rhetoric. As in informal negotiation within the research community, observational claims are embedded in a series of highly standardised formulations of a social kind, which are used to discredit one's opponents and to strengthen one's own assertions. Clearly neither side in this kind of debate is engaged in formal scientific demonstration within a common scientific framework. Both are engaged in informal negotiation and are seeking to substantiate conclusions which are appropriate to their social commitments.

In addition to influencing scientists' definition of the problem, social commitments influence the detailed processes of scientific reasoning. Mazur observes that:

> ... complex technical problems of the state-of-the-art require subtle perceptions of the sort which cannot be easily articulated in explicit form. When it is necessary to make a simplifying assumption, and many are reasonable, which simplifying assumption should be made? When data are lacking on a question, how far may one reasonably extrapolate from data of other sources? How trustworthy is a set of empirical observations? These questions all require judgements for which there are no formalised guides and it is here that experts frequently disagree. (1973, p. 251)

Not only do scientists disagree, but they sometimes introduce non-technical resources from the political setting into their analyses in order to resolve their interpretative difficulties. For instance, Mazur discusses scientists' treatment of the relationship between low-level radiation and the incidence of leukaemia. He shows that several models or interpretations have been formulated which appear to be consistent with the available data. The element of judgement which is essential in choosing between these models seems frequently to be dependent on, or at least associated with, scientists' views about public health policy.

> This theoretical ambiguity has major implications for the technical debate over permissible radiation standards. It should be noted that the 'threshold' model implies that dose levels below the threshold will not harm the population (through leukaemia). The 'linear' model implies that there will be some incidence of leukaemia no matter how low the dose to the population. The two models differ, then, on whether or not there is a 'safe' level of radiation exposure for the population. The ambiguous nature of the dose-effect curve is well recognised by radiation biologists, and many assume the 'linear' model, not necessarily because they consider it true, but because it is the most conservative model for purposes of public safety. (Mazur, 1973, p. 254)

In such a situation, one possible option is that of admitting openly that no clear scientific conclusion is possible. But scientists have not normally acted in this way. One reason for this may be that scientists enter the political context as purveyors of certified knowledge. They have nothing to offer other than the supposed certainties of science; and if they were to present their conclusions as no more than 'plausible guesses' based on uncertain foundations, they would carry little political weight. Moreover, scientists are expected to and are usually inclined to present their conclusions in terms of the formal calculus of science. But in order to do this, scientists have to commit themselves to particular informal judgements which subsequently become hidden from view behind an impersonal terminology and a rhetoric which presents scientific findings as objective representations of the external world.

For instance, it is generally agreed among scientists that there is a connection between exposure to ionising radiation and the occurrence of leukaemia and thyroid cancer. The evidence of a connection between radiation and other forms of cancer, however, is believed by many to be less compelling. There is, therefore, a considerable range of opinion on this issue, each opinion being supported by differing estimates of the reliability of particular experiments and observations, by varying interpretations of the causal processes involved, and so on. In the political context, the important question is whether given levels of radiation constitute a high or low risk to public health. Scientists have responded to this question by calculating the number of cancer cases per year which will be produced by low-level radiation. The figures produced have varied enormously, at least partly because those responsible for these calculations have begun from different basic judgements about the relationship between radiation and the various types of cancer. At one extreme there have been a few vocal critics of the Atomic Energy Commission, who have asserted as a 'scientific law' that all major forms of cancer are produced by radiation and who have consequently calculated the risks of radiation to be very high. At the other end of the spectrum there have been scientists, such as those working for the AEC, claiming that this estimate 'does not square with the facts' and that calculations should begin from the established premise that only leukaemia and thyroid cancer are caused by radiation.

What has happened here and in other instances is that scientists have quantified and applied formal techniques of inference to their informal judgements; they have presented the results as incontrovertible facts; and they have sought to persuade the uncommitted by using the social and technical rhetoric of their professional community. All areas of scientific research are characterised by situations in which the established technical culture permits the formulation of several reasonable alternatives, none of which can be shown conclusively to be more correct than another. It is in making choices between such alternatives, whether at the level of broad definitions of the problem or at the level of detailed analysis, that scientists' political commitments and the pressures of the political context come into play most clearly.

In this section I have tried to show that when scientists enter social contexts outside the research community, such as the wider realm of political activity, they select from and reinterpret their cultural resources, both technical and social, in response to the social context and in accordance with their position in it. This brief and incomplete discussion of the movement of cultural resources out of the research community supplements the prior examination of external influences on the content of science. These two sections complete the preliminary analysis offered in this book by showing that the processes of negotiation and cultural reinterpretation within the research community are linked to similar processes occurring in society at large. They show that scientific knowledge must be seen as one part of a complex movement of cultural resources throughout society—a movement which is mediated through and moulded by the changing pattern of social relationships and by the clash of group interests.

BRIEF SUMMARY AND CONCLUDING REMARKS

I began this book by showing that, because sociologists have customarily regarded scientific knowledge as having a special epistemological status, they have treated the production and legitimation of scientific formulations as a special case within the sociology of knowledge. The content of scientific knowledge has been excluded from sociological analysis because it has been assumed that scientists have found ways of ensuring that their conclusions are determined by the nature of the physical world itself. Once sociologists had accepted that science provided an objective representation of the external world, it appeared to follow that the scientific community had to have certain distinctive characteristics; for it seemed difficult to conceive how objective knowledge could be continually generated by a community which did not have these characteristics. In particular, the scientific research community was seen as necessarily having an intellectually open and universalistic normative structure, as being politically neutral and as operating most effectively in societies which allowed science considerable autonomy.

The standard philosophical view of science, which sociologists of knowledge took for granted in treating science as a special case, furnished a fairly coherent account of scientific observation, the relation between fact and theory, the uniformity of nature and the criteria used to validate scientific knowledge-claims. In Chapter 2 I tried to show that this traditional philosophical analysis was beset by grave difficulties and I tried to sketch the broad outline of an alternative view which grows out of recent philosophical debate. I argued for example, that the principle of the uniformity of nature is best seen, not as an assumption that sociologists themselves have to make about the physical world, but rather as a part of scientists' resources for constructing *their* accounts of that world. I also argued that the factual content of science should not be treated as a culturally unmediated reflection of a stable external world. Fact and theory, observation and presupposition, are inter-related in a complex manner; and the empirical conclusions of science must be seen as interpretative constructions, dependent for their meaning upon and limited by the cultural resources available to a particular social group at a particular point in time. Similarly, general criteria for assessing scientific knowledge-claims cannot be applied universally, independently of social context, as most sociologists have previously assumed. These criteria are always open to varied interpretations and are given meaning in terms of particular scientists' specific intellectual commitments, presuppositions and objectives. In short, I argued that the cognitive/technical resources of scientists are open to continual change of meaning; that there is, therefore, nothing in the physical world which uniquely determines scientists' conclusions; and that consequently it is appropriate for sociologists to look closely at the ways in which scientists construct their accounts of the world and at the ways in which variations in social context influence the formation and acceptance of scientific assertions.

In the third chapter I showed that the longstanding sociological analysis of the normative structure of science was inadequate, quite apart from its inconsistency with the revised philosophical position proposed in Chapter 2. I suggested that what had been taken by sociologists to be a complete set of basic principles specifying proper conduct for scientists engaged in research should be seen as no more than part of a complex social repertoire which scientists use flexibly in the course of negotiating the meaning of their own and their colleagues' actions. In addition, I examined a number of recent case studies which appeared to show that there is no clear separation in science between the negotiation of social meaning and the assessment of knowledge-claims. Both social and cognitive/technical formulations have to be selected and interpreted by participants in particular instances; and both kinds of resource acquire their specific meaning as they are combined in the sequence of informal interaction plus formal demonstration whereby knowledge-claims come to be ratified.

Once we have abandoned the orthodox philosophical view of science, it becomes possible to accept that the social negotiation of knowledge in various kinds of scientific research context is a legitimate, indeed a fundamental and as yet largely unexplored, topic for the sociology of knowledge. But it must never be assumed that the production of knowledge by scientific specialists is entirely divorced from the wider social and cultural context. It is in recognising this that the argument I have advanced here links up with the more traditional concerns of sociologists of knowledge. In the final chapter I have used the case of Darwinian theory to show in detail how cultural resources from society at large can enter into the very form and content of scientific claims as well as, possibly, playing a significant part in their acceptance. Sometimes the connection between science and the wider society is established by means of direct social contact between scientists and outsiders; but it can also arise in a more diffuse way through scientists' ability to select from and reinterpret cultural resources generally available during a specific period to the members of a particular society or a particular social class.

In this last chapter I also considered briefly the implications of my overall thesis for another central concern of the sociology of knowledge, namely, the political actions of knowledge-producers. I suggested that, as one would expect in the light of my previous argument, scientists' knowledge-claims can be affected by their position in a political context and that elements of a political context may become built into scientists' assertions about the natural world. I also tried to show that the growing engagement of scientists in the political arena in no way signals the end of political ideology, as had been widely assumed. I argued instead that scientists' own claim to be politically neutral was itself ideological, in the sense that it constitutes a selective employment and interpretation of the cultural resources available to scientists, in a way which favours the vested interests of their specialised community.

The argument presented in this book and summarised immediately above opens up numerous lines of empirical research and analysis which have previously been largely ignored by sociologists. For instance, once we cease to take it as self-evident that scientific knowledge-claims are assessed by clear-cut, pre-established criteria, it becomes possible to accept that non-technical considerations may systematically influence the allocation of scientific credit. Consequently, it becomes possible to approach the study of social ranking in science in a radically new way and to explore for the first time how far the phenomena of power and domination are present within the research community. This possibility was discussed at the end of the first chapter. In the second chapter, one of the most interesting ideas to emerge was that scientific propositions are not stable in meaning, but are reinterpreted as they move from one social context to another. This process of reinterpretation is at present little understood and is in need of detailed sociological investigation. For example, it would be particularly helpful to have studies of the variations introduced into specific general-purpose formulations in various contexts as well as careful analysis of how these variations were devised to meet the requirements of these contexts.

In Chapter 3, the central theme was that of the social negotiation of scientific knowledge. As I stressed in that chapter, my treatment of this topic left many gaps and, therefore, many avenues for further study. One important point which did become clear was that the evaluative repertoire of scientists is much more complex and extensive than sociologists have previously assumed. Accordingly, if we are to achieve a better understanding of the nature of social negotiation in science, we must have more studies like that of Mitroff which provide wide-ranging and detailed documentation of the moral language of science. But this alone, although essential, is not enough. For we must also find various ways of ascertaining just how the repertoire of social evaluations contributes to the interpretation and acceptance of specific knowledge-claims.

In this final chapter the main theme has been the movement of interpretative resources into and out of the research community. A few political scientists such as Nelkin and Mazur have already begun to study how scientists adapt their specialised knowledge to the pressures of political debate. Such studies are, however, still few in number and I hope that this book will convince some sociologists that this is a potentially fruitful realm of inquiry. But an equally important and related topic, which has been completely ignored by social scientists until very recently, is that of the influences on scientific knowledge originating outside the social networks within which scientific research is carried out. Thus it has become clear in the course of the present chapter that one of the major tasks now facing sociologists of knowledge is to portray the dynamic social processes whereby science absorbs, reinterprets and refurbishes the cultural resources of modern industrial societies. From the new perspective advanced here, science should not be treated as a privileged sociological case and kept separate from other areas of cultural production. Rather, every effort should be made to investigate scientists' debt to the wider society and to delineate the complex connections between cultural production in science and in other areas of social life.

These are just a few of the more obvious topics for further research which grow out of the analysis presented in this book. In addition there are numerous issues which are less easily perceived, yet which require careful examination in the light of this new conception of the sociology of science. Perhaps the most important of these is the question of the relationship between science and technology. From the standard view of science this relationship is relatively unproblematic. For effective technology is seen as a simple by-product of objective knowledge. But if we stress the socially and culturally contingent character of scientific knowledge, we must be prepared to question the widespread assumption that modern technology is on the whole a derivative of basic scientific research and/or to move towards an analysis of the social meaning of technology. I do not intend to pursue this kind of complex issue any further here, however. I hope that many readers will do this for themselves and that they discover in the text various interesting topics and unresolved issues worthy of systematic study that I have not identified explicitly. If this happens, this book will have achieved one of its central aims, namely, that of helping to make the study of science a lively region within the sociology of knowledge.

References

Agassi, J. (1963), 'Towards an historiography of science', *History and Theory*, vol. 2, pp. 1-23.

Barber, B. (1952), *Science and the Social Order* (New York: The Free Press).

Barnes, B. (1974), *Scientific Knowledge and Sociological Theory* (London: Routledge & Kegan Paul).

Barnes, B., and Dolby, R. G. (1970), 'The scientific ethos: a deviant viewpoint', *European Journal of Sociology*, vol. 2, pp. 3-25.

Barnes, B., and Law, J. (1976), 'Whatever should be done with indexical expressions?', *Theory and Society*, vol. 3, pp. 223-37.

Bell, D. (1960), *The End of Ideology* (Glencoe, Ill.: The Free Press).

Ben-David, J. (1968), *Fundamental Research and the Universities* (Paris: OECD).

Ben-David, J. (1971), *The Scientist's Role in Society* (Englewood Cliffs, NJ: Prentice-Hall).

Ben-David, J. (1977), 'Organization, social control, and cognitive change in science', in J. Ben-David and T. N. Clarke (eds), *Culture and Its Creators* (Chicago and London: University of Chicago Press).

Benveniste, G. (1972), *The Politics of Expertise* (Berkeley, Calif.: Glendessary Press).

Berger, P., and Luckmann, T. (1967), *The Social Construction of Reality* (London: Allen Lane).

Blissett, M. (1972), *Politics in Science* (Boston, Mass.: Little, Brown).

Bloor, D. (1976), *Knowledge and Social Imagery* (London: Routledge & Kegan Paul).

Blum, A. (1970), 'The corpus of knowledge as a normative order', in J.C. McKinney and E. A. Tiryakian (eds), *Theoretical Sociology* (New York: Appleton-Century-Crofts).

Blume, S. S. (1974), *Toward a Political Sociology of Science* (New York and London: Collier Macmillan).

Bohm, D. (1965). *The Special Theory of Relativity* (New York: Benjamin).

Böhme, G. (1975), 'The social function of cognitive structures', in K. Knorr, H. Strasser and H. G. Zilian (eds), *Determinants and Controls of Scientific Development* (Dordrecht, Holland: Reidel).

Böhme, G. (1977), 'Cognitive norms, knowledge-interests and the constitution of the scientific object', in E. Mendelsohn, P. Weingart and R. Whitley (eds), *The Social Production of Scientific Knowledge* (Dordrecht, Holland: Reidel).

Böhme, G., Daele, W. van den, and Krohn, W. (1972), 'Alternativen in der Wissenschaft', *Zeitschrift für Soziologie*, vol. 1, pp. 302-16.

Böhme, G., Daele, W. van den, and Weingart, P. (1976), 'Finalization in science', *Social Science Information*, vol. 15, pp. 307-30.

Borger, R., and Seaborn, A. E. (1966), *The Psychology of Learning* (Harmondsworth: Penguin).

Bourdieu, P. (1975), 'The specificity of the scientific field and the social conditions of the progress of reason', *Social Science Information*, vol. 14, pp. 19-47.

Braverman, H. (1974), *Labor and Monopoly Capital* (New York and London: Monthly Review Press).

Brooks, H. (1964), 'The scientific advisor', in R. Gilpin and C. Wright (eds), *Scientists and National Policy Making* (New York: Columbia University Press).

Bruner, J. S. (1974), *Beyond the Information Given* (London: Allen & Unwin).

Brush, S. G. (1967), 'Thermodynamics and history', *The Graduate Journal*, vol. 7, pp. 477-564.

Bunge, M. (1967), 'Technology as applied science', *Technology and Culture*, vol. 8, pp. 329-47.

Burtt, E. A. (1924), *The Metaphysical Foundations of Modern Science* (London: Routledge & Kegan Paul).

Carnap, R. (1939), *Foundations of Logic and Mathematics* (Chicago: University of Chicago Press).

Carnap, R. (1966), *Philosophical Foundations of Physics* (New York: Basic Books).

Chubin, D. E. (1976), 'The conceptualization of scientific specialties', *The Sociological Quarterly*, vol. 17, pp. 448-76.

Cole, J. R., and Cole, S. (1973), *Social Stratification in Science* (Chicago and London: University of Chicago Press).

Collins, H. M. (1974), 'The TEA set: tacit knowledge and scientific networks', *Science Studies*, vol. 4, pp. 165-85.

Collins, H. M. (1975), 'The seven sexes: a study in the sociology of a phenomenon, or the replication of experiment in physics,' *Sociology*, vol. 9, pp. 205-24.

Collins, H. M., and Harrison, R. G. (1975), 'Building a TEA laser: the caprices of communication', *Social Studies of Science*, vol. 5, pp. 441-50.

Collins, H. M., and Pinch, T. J. (1978), 'The construction of the paranormal: nothing unscientific is happening', in R. Wallis (ed.), *Rejected Knowledge: Sociological Review Monograph* (Keele: University of Keele).

Curtis, J. E., and Petras, J. W. (1970), *The Sociology of Knowledge* (New York: Praeger).

Daele, W. van den, Krohn, W., and Weingart, P. (1977), 'The political direction of scientific development', in *The Social Production of Scientific Knowledge* (Dordrecht, Holland: Reidel).

Daniels, G. H. (1967), 'The pure-science ideal and democratic culture', *Science*, vol. 156, pp. 1699-705.

Darwin, C. (1859), *On the Origin of Species* (London: Murray).

De Gré, G. (1955), *Science as a Social Institution* (New York: Random House).

Deutsch, M. (1959), 'Evidence and inference in Nuclear research', in D. Lerner (ed.), *Evidence and Inference* (Glencoe, Ill.: The Free Press).

Duhem, P. (1962), *The Aim and Structure of Physical Theory* (New York: Atheneum).

Durkheim, E. (1915), *The Elementary Forms of the Religious Life* (London: Allen & Unwin).

Durkheim, E. (1938), *The Rules of Sociological Method* (Glencoe, Ill.: The Free Press).

Easlea, B. (1973), *Liberation and the Aims of Science* (London: Chatto & Windus).

Edge, D. O., and Mulkay, M. J. (1976), *Astronomy Transformed* (New York: Wiley Interscience).

Elias, N. (1971), 'Sociology of knowledge: new perspectives', *Sociology*, vol. 5, pp. 149-68, and 355-70.

Elkana, Y. (1974), *The Discovery of the Conservation of Energy* (London: Hutchinson Educational).

Elliot, H. C. (1974), 'Similarities and differences between science and common sense', in R. Turner (ed.), *Ethnomethodology* (Harmondsworth: Penguin) pp. 21-6.

Ezrahi, Y. (1971), 'The political resources of American science', *Science Studies*, vol. 1, pp. 117-33.

Fay, B. (1975), *Social Theory and Political Practice* (London: Allen & Unwin).

Feuer, L. S. (1971), 'The social roots of Einstein's theory of relativity', *Annals of Science*, vol. 27, pp. 277-98, 313-43.

Feyerabend, P. (1975), *Against Method* (London: NLB).

Fisher, R. A. (1936), 'Has Mendel's work been rediscovered?', *Annals of Science*, vol. 1, pp. 115-37.

Forman, P. (1971), 'Weimar culture, causality and quantum theory, 1918-1927: adaptation by German physicists and mathematicians to a hostile intellectual environment', in R. McCormach (ed.), *Historical Studies in the Physical Sciences, No. 3* (Philadelphia Penn.: University of Pennsylvania Press).

Fox, R. (1974), 'The rise and fall of Laplacian Physics', *Historical Studies in the Physical Sciences*, vol. 4, pp. 89-136.

Frake, C. O. (1961), 'The diagnosis of disease among the Subanun of Mindanao', *American Anthropologist*, vol. 63, pp. 113-31

Frank, G. P. (ed), (1961), *The Validation of Scientific Theories* (New York: Collier).

Frankel, E. (1976), 'Corpuscular optics and the wave theory of light: the science and politics of a revolution in physcis', *Social Studies of Science*, vol. 6, pp. 141-84.

Galileo, G. (1953), *A Dialogue Concerning the Two Chief World Systems*, ed. Santillana, G. de (Chicago: University of Chicago Press).

Gaston, J. (1973), *Originality and Competition in Science* (Chicago and London: University of Chicago Press).

Giddens, A. (1978), 'Positivism and its critics', in T. B. Bottomore and R. Nisbet (eds), *A History of Sociological Analysis* (New York: Basic Books).

Gilbert, G. N. (1976a), 'The transformation of research findings into scientific knowledge', *Social Studies of Science*, vol. 6, pp. 281-306.

Gilbert, G. N. (1976b), 'The development of science and scientific knowledge: the case of radar meteor research', in G. Lemaine (ed.), *Perspectives on the Emergence of Scientific Disciplines* (The Hague and Paris: Mouton).

Gilbert, G. N. (1977a), 'Competition, differentiation and careers in science', *Social Science Information*, vol. 16, pp. 103-23.

Gilbert, G. N. (1977b), 'Referencing as persuasion', *Social Studies of Science*, vol. 7, pp. 113-22.

Glass, B. (1953), 'The long neglect of a scientific discovery: Mendel's laws of inheritance' in G. Boas *et al.* (eds), *Studies in Intellectual History* (Baltimore, Maryland: Johns Hopkins Press).

Gouldner, A. W., *The Coming Crisis of Western Sociology* (London: Heinemann).

Grandy, R. E. (1973), *Theories and Observation in Science* (Englewood Cliffs, NJ: Prentice-Hall).

Greenberg, D. S. (1969), *The Politics of American Science* (Harmondsworth: Penguin).

Haberer, J. (1969), *Politics and the Community of Science* (New York: Van Nostrand Rheinhold).

Habermas, J. (1972), *Knowledge and Human Interests* (London: Heinemann).

Hagstrom, W. O. (1965), *The Scientific Community* (New York: Basic Books).

Hanson, N. R. (1965), *Patterns of Discovery* (Cambridge: Cambridge University Press).

Hanson, N. R. (1969), *Perception and Discovery* (San Francisco: Freeman, Cooper).

Harris, E. E. (1970), *Hypothesis and Perception* (London: Allen & Unwin).

Harwood, J. (1976), 'The race-intelligence controversy', *Social Studies of Science*, vol. 6, pp. 369-94, and vol. 7 (1977), pp. 1-30.

Hempel, C. G. (1965), *Aspects of Scientific Explanation* (New York and London: The Free Press).

Hesse, M. (1974), *The Structure of Scientific Inference* (London: Macmillan).

Hessen, B. (1931), 'The social and economic roots of Newton's *Principia*', in N. I. Bukharin *et al.* (eds), *Science at the Crossroads* (2nd ed 1971) (London: Frank Cass).

Hirsch, W. (1961), 'The autonomy of science in totalitarian societies', *Social Forces*, vol. 40, pp. 15-22.

Hoijer, H. (1964), 'Cultural implications of some Navaho linguistic categories', in D. Hymer (ed.), *Language and Culture in Society* (New York and London: Harper & Row), pp. 142-53.

Holton, G. (1973), *Thematic Origins of Scientific Thought* (Cambridge, Mass.: Harvard University Press).

Johnston, R. (1976), 'Contextual knowledge', *Australia and New Zealand Journal of Sociology*, vol. 12, pp. 193-203.

Johnston, R. (1978), 'Goal direction of scientific research', in W. Krohn, E. Layton and P. Weingart (eds), *The Dynamics of Science and Technology*, Sociology of the Sciences Yearbook, Vol. II (Dordrecht, Holland: Reidel).

Johnston, R., and Shepherd, J. (1976), *Science and Rationality* (Manchester: University of Manchester, Department of Liberal Studies in Science, Siscon Pamphlet).

Joravsky, D. (1970), *The Lysenko Affair* (Cambridge, Mass.: Harvard University Press).

Kemp, R. V. (1977), 'Controversy in scientific research and tactics of communication', *Sociological Review*, vol. 25, pp. 515-34.

King, L. R., and Melanson, P. H. (1972), 'Knowledge and politics', *Public Policy*, vol. 20, pp. 82-101.

King, M. D. (1971), 'Reason, tradition and the progressiveness of science', *History and Theory*, vol. 10, pp. 3-32.

Knorr, K., Strasser, H., and Zilian, H. G. (1975), *Determinants and Controls of Scientific Development* (Dordrecht, Holland: Reidel).

Kroeber, A. L. (1944), *Configurations of Culture Growth* (Berkeley, Calif., and Los Angeles: University of California Press).

Kuhn, T. S. (1962), *The Structure of Scientific Revolutions* (Chicago and London: University of Chicago Press) (enlarged ed 1970).

Kuhn, T. S. (1963), 'The function of measurement in modern physical science', in H. Woolf (ed.), *Quantification* (New York: Bobbs-Merrill), pp. 31-61.

Lakatos, I. (1968), 'Criticism and the methodology of scientific research programmes', *Proceedings of the Aristotelian Society*, vol. 69, pp. 149-86.

Lakatos, I. (1970), 'Falsification and the methodology of scientific research programmes', in I. Lakatos and A. Musgrave (eds), *Criticism and the Growth of Knowledge* (Cambridge: Cambridge University Press).

Lakatos, I. (1976), *Proofs and Refutations* (Cambridge: Cambridge University Press).

Lakatos, I., and Musgrave, A. (1970), *Criticism and the Growth of Knowledge* (Cambridge: Cambridge University Press).

Lakoff, S. (1977), 'Scientists, technologists and political power', in I. Spiegel-Rösing and D. J. de Solla Price (eds), *Science, Technology and Society* (London: Sage).

Lane, R. (1966), 'The decline of politics and ideology in a knowledgeable society', *American Sociological Review*, vol. 31, pp. 649-62.

Law, J. (1975), 'Is epistemology redundant? A sociological view', *Philosophy of Social Science*, vol. 5, pp. 317-37.

Law, J. (1976), 'Theories and methods in the sociology of science: an interpretative approach', in G. Lemaine *et al.* (eds), *Perspectives on the Emergence of Scientific Disciplines* (Paris: Mouton, and Chicago: Aldine).

Law J., and French, D. (1974), 'Normative and interpretative sociologies of science', *Sociological Review*, vol. 22, pp. 581-95.

Lemaine, G. *et al.* (eds) (1976), *Perspectives on the Emergence of Scientific Disciplines* (The Hague/Paris: Mouton, and Chicago: Aldine).

Lemaine, G., and Matalon, B. (1969), 'La Lutte pour la vie dans la cité scientifique', *Revue Française de Sociologie*, vol. 10, pp. 139-65.

Lewis, C. I. (1956), *Mind and the World Order* (New York: Dover Publications).

MacKinnon, E. A. (1972), *The Problem of Scientific Realism* (New York: Appleton-Century-Crofts).

MacLeod, R. (1977), 'Changing perspectives in the social history of science', in I. Spiegel-Rösing and D. J. de Solla Price (eds), *Science, Technology and Society* (London: Sage), pp. 149-96.

Mannheim, K. (1936), *Ideology and Utopia* (New York: Harcourt, Brace & World).

Mannheim, K. (1952), *Essays on the Sociology of Knowledge* (London: Routledge & Kegan Paul).

Marcuse, H. (1962), *One-Dimensional Man* (London: Routledge & Kegan Paul).

Marx, K. (1904), *A Contribution to the Critique of Political Economy* (Chicago: C. H. Kerr.)

Marx, K., and Engels, F. (1965), *Manifesto of the Communist Party* (Moscow: Progress Publishers).

Marx, K. (1973), *Grundrisse* (Harmondsworth: Penguin).

Marx, K. (1974), *Economic and Philosophical Manuscripts of 1844* (Moscow: Progress Publishers).

Mason, S. F. (1962), *A History of the Sciences* (New York: Collier).

Mazur, A. (1973), 'Disputes between experts', *Minerva*, vol. 11, pp. 243-62.

McHugh, P. (1971), 'On the failure of positivism', in J. D. Douglas (ed.), *Understanding*

Everyday Life (London: Routledge & Kegan Paul).

Medawar, P. (1969), *The Art of the Soluble* (Harmondsworth: Penguin).

Mendelsohn, E. (1964), 'The emergence of science as a profession in nineteenth-century Europe', in K. Hill (ed.), *The Management of Scientists* (Boston, Mass.: Beacon Press).

Merton, R. K. (1957), *Social Theory and Social Structure* (revised and enlarged ed) (New York: The Free Press).

Merton, R. K. (1970), *Science, Technology and Society in Seventeenth-Century England* (New York: Harper & Row) (originally published in 1938).

Merton, R. K. (1973), *The Sociology of Science* (Chicago and London: University of Chicago Press) (ed. with an introduction by N. W. Storer).

Merton, R. K. (1975), 'Structural analysis in sociology', in P. Blau (ed.), *Approaches to the Study of Social Structure* (New York: The Free Press), pp. 21-52.

Mitroff, I. I. (1974), *The Subjective Side of Science* (Amsterdam: Elsevier).

Mulkay, M. J. (1969), 'Some aspects of cultural growth in the natural sciences', *Social Research*, vol. 36, pp. 22-52.

Mulkay, M. J. (1972), *The Social Process of Innovation* (London: Macmillan).

Mulkay, M. J. (1974), 'Methodology in the sociology of science', *Social Science Information*, vol. 13, pp. 107-19.

Mulkay, M. J. (1976a), 'Norms and ideology in science', *Social Science Information*, vol. 15, pp. 637-56.

Mulkay, M. J. (1976b), 'The mediating role of the scientific elite', *Social Studies of Science*, vol. 6, pp. 445-70.

Mulkay, M. J. (1977a), 'Connections between the quantitative history of science, the social history of science and the sociology of science', *Proceedings of the International Seminar on Science Studies* (Helsinki: Academy of Finland), pp. 54-76.

Mulkay, M. J. (1977b), 'Sociology of the scientific research community', in I. Spiegel-Rösing and D. J. de Solla Price (eds), *Science, Technology and Society* (London: Sage), pp. 93-148.

Mulkay, M. J., Gilbert, G. N., and Woolgar, W. (1975), 'Problem areas and research networks in science', *Sociology*, vol. 9, pp. 187-203.

Mulkay, M. J. and Williams, A. T. (1971), 'A sociological study of a physics department', *British Journal of Sociology*, vol. 22, pp. 68-82.

Musson, A. E. and Robinson, E. (1969), *Science and Technology in the Industrial Revolution* (Manchester: Manchester University Press).

Nagel, E. (1961), *The Structure of Science* (London: Routledge & Kegan Paul).

Nagel, E., Bromberger, S., and Grünbaum, A. (1971), *Observation and Theory in Science* (Baltimore and London: Johns Hopkins Press).

Nelkin, D. (1971), 'Scientists in an environmental controversy', *Science Studies*, vol. 1, pp. 245-61.

Nelkin, D. (1975), 'The political impact of technical expertise', *Social Studies of Science*, vol. 5, pp. 35-54.

Nelkin, D. (1977), 'Scientists and professional responsibility: the experience of American ecologists', *Social Studies of Science*, vol. 7, pp. 75-95.

O'Neil, W. N. (1969), *Fact and Theory* (Sydney: Sydney University Press, and London: Methuen).

Outhwaite, W. (1975), *Understanding Social Life* (London: Allen & Unwin).

Pinch. T. J. (1977), 'What does a proof do if it does not prove?', in E. Mendelsohn, P. Weingart and R. Whitley (eds), *The Social Production of Scientific Knowledge* (Dordrecht, Holland: Reidel).

Polanyi, M. (1951), *The Logic of Liberty* (London: Routledge & Kegan Paul).

Polyani, M. (1958), *Personal Knowledge* (London: Routledge & Kegan Paul).

Polanyi, M. (1969), *Knowing and Being* (London: Routledge & Kegan Paul).

Popper, K. R. (1959), *The Logic of Scientific Discovery* (New York: Harper & Row).

Popper, K. R. (1963), *Conjectures and Refutations* (London: Routledge & Kegan Paul).

Price, D. J. de Solla (1963), *Big Science, Little Science* (New York: Columbia University Press).

Radnitzky, G. (1974), 'From logic of science to theory of research', *Communication and Cognition*, vol. 7, pp. 61-124.

Ravetz, J. R. (1971), *Scientific Knowledge and Its Social Problems* (Oxford: Clarendon Press).

Robbins, D., and Johnston, R. (1976), 'The role of cognitive and occupational differentiation in scientific controversies', *Social Studies of Science*, vol. 6, pp. 349-68.

Rose, H., and Rose, S. (1976) (eds), *The Political Economy of Science* (London: Macmillan).

Rudwick, M. J. (1976), 'The emergence of a visual language for geological science', *History of Science*, vol. 14, pp. 149-95.

Ryle, G. (1949), *The Concept of Mind* (London: Hutchinson).

Sandow, A. (1938), 'Social factors in the origin of Darwinism', *The Quarterly Review of Biology*, vol. 13, pp. 315-26.

Scheffler, I. (1963), *The Anatomy of Inquiry* (New York: Knopf).

Scheffler, I. (1967), *Science and Subjectivity* (New York: Bobbs-Merrill).

Schon, D. A. (1963), *Displacement of Concepts* (London: Tavistock).

Sklair, L. (1973), *Organized Knowledge* (London: Hart-Davis, MacGibbon).

Slack, J. (1972), 'Class struggle among the molecules', in T. Pateman (ed.), *Countercourse* (Harmondsworth: Penguin).

Smart, J. C. (1973), 'The reality of theoretical entities', in R. E. Grandy (ed.), *Theories and Observation in Science* (Englewood Cliffs, NJ: Prentice-Hall).

Stark, W. (1958), *The Sociology of Knowledge* (London: Routledge & Kegan Paul).

Storer, N. W. (1966), *The Social System of Science* (New York: Holt, Rinehart & Winston).

Storer, N. W. (1973), Introduction to Merton, R. K., *The Sociology of Science* (Chicago and London: University of Chicago Press).

Strawson, P. F. (1959), *Individuals* (London: Methuen).

Sullivan, D., White, D. H., and Barboni, E. J. (1977), 'Co-citation analyses of science', *Social Studies of Science*, vol. 7, pp. 223-40.

Thackray, A. (1970), 'Science and technology in the industrial revolution', *History of Science*, vol. 9, pp. 76-89.

Tobey, R. C. (1971), *The American Ideology of National Science* (Pittsburgh, Pa.: University of Pittsburgh Press).

Toulmin, N. J. (1968), *Foresight and Understanding* (New York: Harper & Row).

Vig, N. J. (1968), *Science and Technology in British Politics* (London: Pergamon).

Vorzimmer, P. J. (1969), 'Darwin's questions about the breeding of animals (1839)', *Journal of the History of Biology*, vol. 2, pp. 269-81.

Webster, A. J. (1978), 'Scientific controversy and socio-cognitive metonymy: the case of acupuncture', in R. Wallis (ed.), *Rejected Knowledge: Sociological Review Monograph* (Keele: University of Keele).

Weingart, P. (1970), *Die Amerikanische Wissenschaftslobby* (Dusseldorf: Bertelsmann Universitätsverlag).

Weingart, P. (1974), *Wissenschaftssociologie* (Frankfurt am Main: Atheneum Fischer Taschenbuch Verlag).

Weingart, P. (1974), 'On a sociological theory of scientific change', in R. Whitley (ed.), *Social Processes of Scientific Development* (London: Routledge & Kegan Paul).

Whitley, R. (1972), 'Black-boxism and the sociology of science', in P. Halmos (ed.), *Sociological Review Monograph* (Keele: University of Keele), pp. 61-92.

Whitley, R. (ed.) (1974), *Social Processes of Scientific Development* (London: Routledge & Kegan Paul).

Whitley, R. (1975), 'Components of scientific activities, their characteristics and institutionalisation in specialties and research areas', in K. Knorr, H. Strasser and H. G. Zilian (eds), *Determinants and Controls of Scientific Development* (Dordrecht, Holland: Reidel).

Wittgenstein, L. (1953), *Philosophical Investigations* (Oxford: Blackwell).

Woolgar, S. (1976a), 'Writing an intellectual history of scientific development: the use of discovery accounts', *Social Studies of Science*, vol. 6, pp. 395-422.

Woolgar, S. (1976b), 'The Identification and definition of scientific collectivities', in G. Lemaine *et al.* (eds), *Perspectives on the Emergence of Scientific Disciplines* (Paris: Mouton, and Chicago: Aldine).

Wynne, B. (1976), 'C. G. Barkla and the J phenomenon: a case study of the treatment of deviance in physics', *Social Studies of Science*, vol. 6, pp. 307-47.

Young, R. M. (1969), 'Malthus and the evolutionists: the common context of biological and social theory', *Past and Present* vol. 43, pp. 109-45.

Young, R. M. (1971a), 'Darwin's metaphor: does Nature select?', *The Monist*, vol. 55, pp. 442-503.

Young, R. M. (1971b), 'Evolutionary biology and ideology—then and now', *Science Studies*, vol. 1, pp. 177-206.

Young, R. M. (1973), 'The historiographic and ideological contexts of the nineteenth-century debate on Man's place in Nature', in M. Teich and R. M. Young (eds), *Changing Perspectives in the History of Science* (London: Heinemann).

Index